St. Louis Community College

Library

5801 Wilson Avenue
St. Louis, Missouri 63110

Computer Aided
Structural Design

Computer Aided Structural Design

D. Clarke

Lecturer in Civil Engineering
University of Strathclyde Glasgow

A Wiley–Interscience Publication

JOHN WILEY & SONS

Chichester · New York · Brisbane · Toronto

Library of Congress Cataloging in Publication Data:

Clarke, Dennis.
  Computer aided structural design.

  'A Wiley–Interscience publication.'
  Includes bibliographical references and index.
  1.   Structural design—Data processing.
2.   Structural design—Mathematical models.   I.   Title.
TA658.2.C53   624'.1771 0285   78-1511

ISBN 0 471 99641 6

Photosetting by Thomson Press (India) Limited, New Delhi
and printed in Great Britain by The Pitman Press, Bath, Avon.

Preface

This book is about the programming of structural design problems for solution by electronic digital computers. For the sake of simplicity I have restricted design considerations to those of analysing a structure (the form of which it is assumed has already been decided elsewhere) and proportioning its component parts. The book is primarily written for students and structural engineers who are already familiar with manual design methods; because the book is almost wholly concerned with computer applications of known design techniques the presentation of design theory is limited and given only when it is needed to illuminate a programming problem.

The book has a threefold purpose. The first is to demonstrate that the transition from manual to computer aided design is a simpler one than those presently without computing experience probably realize. The second is to discuss the ways in which a variety of different structural design problems can be solved with the aid of a computer. And the last is to show how these methods may be translated into a simple computer language.

Whilst it is true that relatively few designers will ever play an active role in programming itself, increasing numbers will inevitably be drawn into the general sphere of computing; some will have to advise professional programmers of the needs and problems peculiar to structural design, and many others will make increasing use of existing programs in the course of their work. Whether or not it is the reader's intention of becoming seriously involved in computer programming, the experience to be gained from examining typical programs and studying detailed descriptions of their structure should give him the confidence to decide for himself what can reasonably be expected of computer programs in the way of design methods and the quality of results.

The first chapter in this book sketches in a background against which the basic ideas involved in computer aided design can be seen. Here the reader is introduced to different aspects of computing—a simple computer language, how to go about writing a program, the use of a control language to 'converse' with a computer, and a general description of two major design approaches. As for the ways in which computer systems actually work, I do not believe that the normal user's knowledge of this needs to extend beyond that of knowing the correct program format and the sequence of teleprinter keyboard messages which will initiate the running of a program. This kind of information is readily available to the user of a particular system.

The remaining chapters are devoted to ways of solving specific design problems. Chapter 2 describes two methods of designing reinforced concrete slabs. The bending requirements of prestressed concrete beams are considered in Chapter 3. Chapters 4, 5 and 6 are concerned with methods of designing continuous frames in reinforced concrete and structural steelwork.

Whilst it is not essential to read the text in the precise order in which it is presented, inevitably it will be found that some features which are common to more than one kind of design problem are considered in more detail the first time that they are met with. However, the reader should find that, taken together in groups, Chapters (1 and 2), (1 and 3), (1, 4 and 5) and (1, 4 and 6) are reasonably self-contained parcels of information.

The greatest problems met with in design programming are more likely to arise out of the nature of design itself than from an unfamiliarity with computer languages and operating procedures. Computing is largely a matter of applying rigid rules which are not open to more than one interpretation. But when faced with the problem of translating a design concept into programming terms the designer may well find himself reappraising his own attitudes towards design, possibly to the extent of appreciating for the first time the true effects of the decisions he has become used to making. Well established design procedures (e.g. the evaluation of design force actions, the adoption of sections from a standard list, the specification of covers to reinforcement, etc.) are straightforward matters to define and translate into programming terms. Where difficulties are most likely to arise is when attempting to program the often intuitive process of making a decision when confronted by a number of conflicting options.

The object of a programming exercise is to establish a path connecting basic arithmetical steps; the direction of the path dictates the order in which the steps are executed and hence the outcome of the calculation. Some groups of arithmetical steps define procedures that are common to a variety of design situations, e.g. the search for a tabulated value, the replacement of a differential equation with its finite-difference equivalent, the devising of iterative calculations and establishing criteria for convergence, etc. When a computing technique has been developed to meet one situation it will often be applicable to others with little modification. These common factors combine to give design programs which have many similar features.

If it is possible to design manually then it is equally possible to create a computer aided design tool which will produce the same result in a fraction of the time. But, whilst it is a worthwhile goal to relieve the designer of tedious arithmetic, this attitude does no more than just begin to exploit the potential offered by modern computers. The opportunity of executing otherwise impossible arithmetical feats in finite time should encourage the development of programs which give a more realistic assessment of the behaviour of structures fabricated from reinforced concrete and other materials for which the assumption of linear behaviour is, at the best, a facile expedient.

There is no single 'best' solution to a design problem. The client's requirements, building regulations, structural form, materials and overall cost all contribute to what must inevitably be a compromise; but in this situation the arbiter of what is indeed acceptable should continue to be the designer and not the programmer. Even though the computer is no substitute for a well-founded design intuition a useful byproduct of its use is that a designer may build up a fund of experience more rapidly than hitherto. Freed from time consuming calculations a number of different schemes can be processed in a short time, thus allowing the designer to devote himself to planning, evaluating results and decision making.

Whether or not these matters should also be left to the programmers is both arguable and outside the scope of this book, but having raised the issue it deserves some comment. In my view planning can never be a wholly logical process; in part it should be the product of the planner's intuition. When given a set of correct assumptions to work with a computer can always furnish hard facts; it can always answer 'yes' or 'no' but it cannot introduce a random 'in spite of' or a 'maybe'—the vagaries of human illogic which occasionally produce something more than just another plan. Computer aided planning should therefore be a carefully considered process of man–machine interaction and not be wholly the end-product of a master program.

Such attitudes affect one's view of computer aided design as well. It is entirely possible to write master design programs which are capable of producing a solution automatically with no intervention from the program user other than his initial specification of the design parameters. But these are single solutions—the unique outcome of monolithic programmed constraints. They are optimum solutions only in so far as they meet the *programmer*'s criteria. In practice optimization is largely an intuitive exercise which is often better served by programs which give the designer himself the opportunity of finding a route to the solution.

The use of any design aid, whether it be a computer program or a set of design curves, carries with it the responsibilities of choosing an aid which is suitable for the purpose, applying it in its correct context and the most important of all, reviewing the results critically. Only by cultivating an attitude of healthy scepticism will the computer aided design tool be kept in its correct perspective. It should be accepted for what it is—a powerful *aid* to design; but it must not be allowed to override the designer's sense of what is right.

D. CLARKE

Contents

Chapter 1

An Introduction to Computer Programming

1.1 What is a Computer Program?

A program is a sequence of coded instructions which control the operation of a computer. These instructions must be presented to the computer as a logical succession of events because it is the nature of most computer systems that they are only able to respond to commands in the precise order in which they are presented. And whilst their overwhelming advantage over manual methods of calculation is the speed with which they are able to accomplish their task, it is indeed a truism that a computer which is fed with rubbish will output rubbish.

Amongst other things then, programming is about the order in which things are done. It follows from this that it is largely impossible to write a successful program without first having an intimate knowledge of the problem to be solved and the form of its solution. Given that knowledge, program development is often helped if a flow diagram is constructed. According to the complexity of the problem this can be produced in various degrees of detail to outline the order in which it is intended to carry out the solution. It will also serve as a reminder of the route that the solution will take whilst the problem is being translated into a state to which the computer will respond—a computer language.

In common with all languages, those peculiar to computers have their own vocabulary and syntax. Man-to-man communication does not necessarily break down if words are misspelt, or if a sentence is ended with a preposition, or an infinitive is split. Between man and machine, however, correct spelling and syntax are of paramount importance. If the rules of language are not obeyed implicitly then the machine will not 'comprehend' and this will result in it either refusing to work on the problem or producing erroneous results.

It is a rare occurrence for a program of any complexity to work perfectly the first time that it is run on a computer. But when all the programming errors have been tracked down and the results of the calculation appear, neatly typed in well ordered rows under informative headings—then take care. Such results have a mesmeric quality and a spurious authority out of all proportion to similar ones which were the product of sliderule manipulation. It is well to remember that the results of a computer calculation are no better than the assumptions on which they were based.

A program should not be considered as complete until it has been carefully documented. The task of modifying the program in the future is a much simpler one if a full program description is available for reference. Of even more importance is the fact that others may wish to use the program and it should be possible for them to verify for themselves the fundamental assumptions on which it is founded. Any other course could lead to a serious propagation of errors.

1.2 BASIC—A Simple Computer Language

1.2.1 Introduction

Computer jargon is studded with acronyms; occasionally they are also meaningful descriptive mnemonics. BASIC is one such acronym. It evolves from the title 'Beginner's All-purpose Symbolic Instruction Code'.

The information concerning BASIC language which is given in this chapter covers only those aspects which are used in the accompanying programs. For a full description of the language and its potential the reader should consult Ref. 1. (See the list of references at the end of the chapter.)

1.2.2 Statements and Line Numbers

A program consists of a number of *lines*. Each line contains one *statement* which represents an instruction to the computer to execute a particular type of operation. The nature of the operation will be governed by the content of the statement. The meaning of the statement will usually be evident from the English language word by which it is prefaced, e.g. PRINT, GOTO, LET, etc. A number assigned to each line by the programmer determines the order in which the statements are obeyed. When writing a program it is advisable to leave a gap between the numbers of consecutive statements to allow further lines to be inserted at a later stage in program development should this prove to be necessary.

1.2.3 Variables

Within a manual calculation either numbers themselves, or symbols representing numbers, may be manipulated. In just the same way a computer may be instructed either to sum two known quantities, e.g. (3.2 + 5.76), or the contents of two variables, e.g. (V2 + C7). In the first case the result will always be 8.96, but in the second case the solution will depend upon the values currently assigned to the variables we have called V2 and C7. Because it is possible to change the content of variables as a calculation progresses, it is therefore possible to execute the same type of calculation any number of times, for different sets of parameters. The essence of computing probably lies in keeping track of what each variable represents and the numerical value it holds.

In BASIC variables may be called by any upper case letter or an upper case letter followed by a single numeral. Thus A, D, F, U0, Z6 and A4 would all be recognized by the computer and treated as variables.

1.2.4 Numbers

A number must be presented to the computer in decimal form. It may either be positive or negative and up to eleven digits in length. Thus 22, 4.3 and -8.6531824776 are acceptable whilst $22/7$ and $\sqrt{3}$ are not. At times it is advantageous to use the form $YE \pm x$ which means 'raise the number Y to the xth power of 10'. Thus 0.00000077864 can be represented by $77864E - 11$.

1.2.5 Arithmetical Operations and Mathematical Functions

The five arithmetical operations, addition, subtraction, multiplication, division and exponentiation, are denoted in a printed program by the symbols $+, -, *, /,$ and $\uparrow$. A mathematical expression is formed by linking together two or more variables with these symbols in the usual way. Thus $(V1 + C3)*A4$ is interpreted as meaning 'add the content of the variable called V1 to the content of C3 and multiply the result by A4'. The result would be quite different to that given by $V1 + C3*A4$. The reason for this is that evaluation takes place from left to right, subject to the precedence rule that exponentiation has precedence over multiplication and division, which in turn has precedence over addition and subtraction. The precedence rule may, however, be overridden by the use of brackets. Generally it is better to insert a superfluity of brackets to ensure the correct interpretation of a mathematical expression than to risk an error by putting in too few.

BASIC language allows nine different mathematical functions to be used when devising mathematical expressions. Of this number, the five used in the accompanying programs are:

SIN (X) Find the sine of X
COS (X) Find the cosine of X
ATN (X) Find the arctangent of X
ABS (X) Find the absolute value of X, i.e. $|X|$
SQR (X) Find the square root of X, i.e. $\sqrt{X}$

(where in each of the above cases X is a number, or an angle in radians, as appropriate).
One or more of these functions may be used in a single mathematical expression. Thus SIN(Z4) and ATN(SQR(ABS(X2 − P3))) are both acceptable. The second of these two expressions means 'find the angle in radians whose tangent

is the square root of the absolute value of the difference between the contents of the variables X2 and P3'.

1.2.6 The LET Statement

This statement is the means whereby the value of an arithmetical expression is transferred to a single variable. It takes the form:

(Line number) LET (Variable) = (The value of an arithmetical expression).

For example:

510 LET F7 = ATN(SQR(ABS(X2 − P3))).

The result of the computation which is carried out on the right hand side of the expression is transferred to F7, and during this exchange the contents of the variables X2 and P3 remain unaltered. An exception to this is where we have a LET statement of the form:

580 LET F7 = F7 + A9

This means 'let the content of the variable called F7 now equal the sum of its *previous* value and the content of A9'.

At an early stage in the development of the accompanying programs it was found by chance that the LET statement is still obeyed even when the command itself is omitted. For this reason the word 'LET' has been omitted from many of the statements which, by the rules of syntax, should formally be preceded by it. Thus the following statements:

440 LET P3 = A5 + Z6
and 440 P3 = A5 + Z6

should be taken as being equivalent to one another.

1.2.7 Arrays and their Use—The DIM Statement

Whether design is a manual or a computer aided process repeated reference must be made to tabulated information—bar areas, section properties, allowable stresses, etc. Furthermore, it is often necessary to refer back to the results given by intermediate steps at a later stage in the overall design process. Any numerical information, whether or not it is of a permanent nature, but which it is convenient to treat as a *set*, is stored together in one area of the computer's memory. A set of related information is called an *array*. Regardless of the physical form that an array actually takes inside the computer, for the purposes of using its information the array may be treated exactly as if it were a bank of pigeon holes. The information in any array element (i.e. pigeon hole) may be retrieved, inspected or modified by specifying the array name and the location of the element in the array.

The array itself is identified by a single upper case letter followed by a pair of brackets, e.g. M( ). It may consist of a single row of elements or a number of rows; the former is a one-dimensional array, the latter two-dimensional. The type of array is recognized once an element has been specified. A one-dimensional array element is specified by its location in the single row. Thus D(7) represents the content of the 7th element in an array called D( ). In a similar way E(4, 2) represents the content of an element in a two-dimensional array called E( ) which is situated at the intersection of the 4th row and the 2nd column of elements. More generally the row and column locations may be specified by integer variables. Thus D(H3) and E(V6, D2) are acceptable descriptions if H3, V6 and D2 have been assigned integer values.

An array element is handled in the same way as an ordinary variable. Thus:

$$320 \text{ LET } U(I,J) = R5 + V6$$
$$330 \text{ LET } V7 = U(P2,C5) + L4$$
$$\text{and } 340 \text{ LET } P(X,Y1) = Q(I,J)/T(L4,N6)$$

are all acceptable statements.

If a one-dimensional array has ten or less elements, or a two-dimensional array has ten or less rows of not more than ten elements, then storage space for the array is automatically allocated by the computer. If one or both of the array dimensions is greater than ten then the computer must be instructed to allocate the necessary space before operations using that array are attempted. In this situation the DIM statement is used. It takes the following form:

$$120 \text{ DIM } C(4,27)$$

This will instruct the computer that an array called C( ) is to be used and that it will have 4 rows and 27 columns of elements, i.e. a total of $4 \times 27 = 108$ elements.

1.2.8 Jumps and Loops

1.2.8.1 The GOTO and IF–THEN Statements

It is often the case that the order in which design operations need to be executed depends upon some factor which is only revealed during the design process itself. This kind of situation demands that the calculation has the capacity to 'jump' from one part of the program to another without obeying the intervening instructions. The GOTO and IF–THEN statements meet this requirement.

The GOTO statement is unconditional and takes the form:

$$150 \text{ GOTO } 210$$

Thus whenever the statement at line 150 is obeyed then any operations defined by lines numbered from 151 to 209 inclusive will be omitted.

In contrast the IF–THEN statement is a conditional jump which takes the form:

240 IF (condition to be satisfied) THEN 110

The 'condition to be satisfied' will rest with a comparison between the present content of a variable and either a numerical quantity or the outcome of a complex algebraic expression. Thus the statement:

240 IF V7 < = (A2 + D9) ↑2/SQR(M) THEN 110

calls for a jump to line 110 if at the time that line 240 is met the content of the variable V7 is less or equal to the value of the right hand side of the expression; otherwise the jump is ignored.

1.2.8.2 The FOR and NEXT Statements

The FOR and NEXT statements are used to embrace a section of program which is to be obeyed a fixed number of times. This operation is called a *loop*. When used in the accompanying programs it always takes the form:

YYY FOR Variable = Constant(1) TO Constant(2) STEP Constant(3)
(Intervening statements)
ZZZ NEXT Value of variable

where YYY and ZZZ are line numbers.

The 'FOR Variable = Constant (1) TO Constant (2)' part of statement YYY specifies the range of values over which the operation is to be carried out; 'STEP Constant (3)' defines the incremental change in the control variable, and hence its intermediate values. Statement ZZZ advances the value of the control variable and provides an exit from the loop when the specified range of values has been covered.

One example of the use of these statements is:

320 FOR I = 4 TO 16 STEP 4
(Intervening statements)
380 NEXT I

This would execute the operation defined by lines 321 to 379 a total of four times, i.e. for values of the control variable equal to 4, 8, 12 and 16.

When the interval between consecutive values of the control variable is 1 the magnitude of the step may be omitted from the FOR statement. Thus:

390 FOR I = 3 TO 8

means that the control variable will be advanced by 1 between boundary values of 3 and 8.

So far we have only considered loops which are defined by the values of a single control variable. Frequently two or more control variables are needed to define the scope of an operation. Consider the following procedure:

```
185 FOR I = 1 TO S
190 FOR J = 1 TO B
195 IF Z(I,J) < = V5 THEN 210
200 LET T6 = Z(I,J)
205 GOTO 220
210 NEXT J
215 NEXT I
220 . . . . . . . . . . . . . etc.
```

The purpose of these statements is to search an array called Z( ), which has S rows and B columns (it is assumed that the values of S and B have already been defined at an earlier stage in the program), for the first element whose content is greater than the value of variable V5. On locating this element its content is to be copied into a variable called T6 and the search discontinued.

Two pairs of FOR and NEXT statements have been nested to give a loop within a loop. The outer loop (defined by lines 185 and 215) causes the control variable I to take successive integer values from 1 to S. The inner loop (defined by lines 190 and 210) is activated for *each* new value of I. This loop advances the control variable J from 1 to B in steps of 1. Since the position of an array element is fully defined by its I (row) and J (column) values then the combined effect of these nested loops is to define each element of each row in turn. The conditional jump at line 195 compares the content of Z(I,J) with that of V5. If Z(I,J) is less or equal to V5 then the value of J is advanced by 1 and the next array element is considered when the jump to line 210 is obeyed. Otherwise, if the content of Z(I,J) is greater than V5 then its value is copied into T6 (see line 200) and the unconditional jump at line 205 effects an exit from the procedure. It should be noted that even though the two FOR statements implied a total of S*B operations it was possible to suspend this activity, once the relevant conditions had been satisfied, by introducing a jump to an area outside the procedure.

1.2.9 Putting Information into the Computer

1.2.9.1 Introduction

A structural design program employs two categories of numerical information. The first is information which is common to all the problems the program is competent to handle; this includes section and material properties, allowable stresses, etc. The second category is concerned with information which defines the problem to be solved, e.g. the geometry of the structure and its loading.

The first type of information is most easily dealt with by building it into the program in a monolithic fashion (see Section 1.2.9.2). In contrast with this it is expedient to present the parameters of an individual problem to the computer at the time a solution is required (see Section 1.2.9.3).

1.2.9.2 The READ and DATA Statements

These statements, positioned at a point in the program before the stage at which the information is required by the calculation, can take the following form:

    200 READ C1, N8, L4
    210 DATA 5, 32.6, 0.3347

Their effect is to make the variables C1, N8 and L4 take the values 5, 32.6 and 0.3347 respectively. The values of these variables would then remain unchanged until either the DATA statement was modified or later instructions in the program changed their values. Whilst it is a simple matter to modify a DATA statement when editing a program, this cannot be done during a program run. For this reason data which is handled by DATA statements become a monolithic part of the program. The main advantage of using READ and DATA statements is that information which is unlikely to require constant modification need only be typed (and checked) once.

1.2.9.3 The INPUT Statement

During a program run the statement:

    240 INPUT L2, W5

would cause the computation to halt and the teleprinter to type a question mark. This represents a request by the computer for information and the computation could not progress further until the designer had typed in two numbers representing the values he had chosen for the variables L2 and W5. It is advisable to incorporate a PRINT statement at this stage in the program to remind the designer of what information is required when the question mark appears. Usually an INPUT statement is preceded by a PRINT state-ment of the following type:

    220 PRINT "YOUR VALUES OF SPAN (M) AND LOAD (KN)
         ARE"
    230 INPUT L2, W5

When the computer reached line 220 the message within the parentheses would be typed out, together with a question mark on the next line. (Had line 220 been terminated by a semicolon then the question mark would have been typed on the same line as the text and immediately after it.) Computation would then resume when the machine had received the two pieces of information.

1.2.10 Getting Information out of the Computer—The PRINT Statement

One example of a PRINT statement was given in Section 1.2.9.3. There it was said that anything appearing within parentheses after the word PRINT would be output by the teleprinter at the time that statement was obeyed. This facility allows relevant text to be output and is particularly useful in providing descriptive headings to results.

The PRINT statement is also the vehicle whereby the numerical solution is conveyed to the designer. A statement of the form:

240 PRINT Q2

will cause the numerical value of the variable Q2 to be output. A more informative variant of this statement is to combine both text and variable output in the following way:

240 PRINT "SHEAR FORCE = " Q2 "KN"

The statement:

250 PRINT

introduces one new line, without printed output, and is useful in providing vertical spaces in between separate blocks of information.

1.2.11 The END Statement

To terminate a program the last statement in it will always be of the form:

720 END

1.3 How to Write a Program

1.3.1 Introduction

A programmer can reasonably expect the computer to do what he has told it to do. But because of errors in logic this will not necessarily be what he *thinks* he has told it to do. The effect of such errors will be contained and more easily identified if the program is built up in a number of stages so that its state can be checked at intervals before it is expanded further.

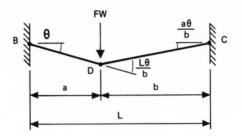

Figure 1.1 Beam collapse mechanism

The complexity of the problem to be solved is usually an indication of the effort needed to develop a computer program for its solution. This effort is usually most economically expended if the program is written in general terms so that it can process the solutions to a wide variety of problems taken from the same general field.

The short program which will be discussed in this section illustrates the application of most of the BASIC language facilities which were described in Section 1.2. This program, called PD1, was devised to choose from a short list of structural steel sections the one required for a single span beam (see Fig. 1.1) carrying a transverse point load which may act at any position in the span. Adequate lateral support is assumed to allow the beam to collapse by the formation of a plastic hinge mechanism. Each end of the span may either be fully fixed or simply supported.

1.3.2 Outline of Steel Beam Collapse Theory

A beam BC, of span L metres, is shown in Fig. 1.1. It is subjected to a collapse point load of FW where F is the Load Factor and W the Working Load in kN. The load is applied at a point D which is distant 'a' from the support B, and 'b' ($= L - a$) from the support C. Assuming full fixity at B and C the beam will collapse by the formation of a three-hinge mechanism. If the hinge at B rotates through a small angle θ then the rotations of the hinges at C and D will be a θ/b and $L\theta/b$ respectively.

Equating the internal work done by the plastic hinges in rotating to the external work done by the moving load gives:

$$M_p X\theta + M_p L\theta/b + M_p Ya\theta/b = FWa\theta \qquad (1.1)$$

where M_p is the plastic moment of resistance of the beam. X and Y are coefficients which can take the value of 0 or 1. By their use the support hinge terms can be suppressed or included according to whether the relevant support is a simple one or fully fixed. In this way a single expression can be used to define the M_p value for a span with four possible sets of support conditions.

From equation (1.1) we get:

$$M_p = FWa/(X + L/b + Ya/b) \text{ kNm} \tag{1.2}$$

and the Plastic Modulus S is given by:

$$S = M_p \, 10^6/f_y \text{ mm}^3 \tag{1.3}$$

where f_y is the yield stress in N/mm^2.

1.3.3 The Flow Diagram

1.3.3.1 Introduction

A useful preliminary exercise, which can be carried out before even attempting to construct a flow diagram, is to compile a list of the basic stages needed in the solution to the problem in their correct order. The list can then serve either as a pattern on which to base a flow diagram or be simply a reminder of the program route whilst it is being written. Unless the route to a solution can be visualised in sufficiently fine detail then at some stage in the programming process flow diagrams will be found to be a useful aid.

A programmer often finds it expedient to begin by developing some other item than the first basic stage which appears on his list. This serves as a nucleus from which the whole program can then expand. If at some point the logic of the program becomes vague then the situation may be reassessed by summarizing the current state of affairs in flow diagram form. This diagram may even be developed past the present stage to show the next steps in the solution. In this way the flow diagram and program development often proceed together.

Although Program PD1 is a trivial one in terms of programming effort it does reflect the fundamental requirements of any design program which, like the whole of Gaul, is divided into three parts:
1. Input the design parameters;
2. Design the structure;
3. Output the results.
These requirements have been expanded to give the flow diagram in its final form as shown in Fig. 1.2.

Blocks of program comprising groups of statements which for the purpose of program development may be divorced from the remainder of the program, are shown in boxes. These boxes are labelled in sequence with letters from A to K. Each box is connected to its neighbours (and in some flow diagrams to others quite remotely situated) by lines along which arrows show the route to be taken. When interpreting a flow diagram the route will always run in the direction of the arrows, never against. Some of the boxes are rectangular, others hexagonal. The former contain instructions to carry out specific operations; the latter represent stages in the program when decisions are made which influence the course of the calculation.

Depending upon the complexity of the problem, flow diagrams can be

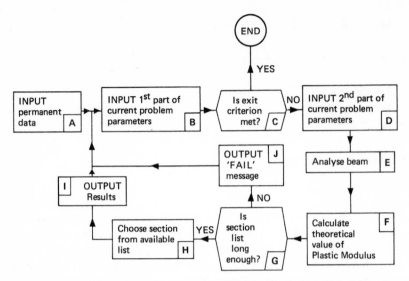

Figure 1.2 Flow diagram for program PD1 (plastic design of steel beam)

produced which reflect various levels of detail. The general flow diagram of the overall problem may serve in developing most of the program. However, individual boxes in the general flow diagram could represent complex situations which in their own right require more detailed flow diagrams for their clarification.

1.3.3.2 The Flow Diagram for Program PD1

In Fig. 1.2 Block (A) together with Blocks (B) and (D) are concerned with the input of information. The different kinds of design data and the reasons for presenting them to the computer in different ways were discussed in Section 1.2.9.1. The permanent data in this design program consists of some properties of four Universal Beam sections; they are input at Block (A) to become an integral part of the program. The designer himself inputs the parameters of the current problem at Blocks (B) and (D). These parameters consist of the beam span and working load, the support conditions, load factor and steel yield stress.

The design calculation takes place at Blocks (E), (F), (G) and (H). The purpose of Block (E) is to calculate the required plastic moment of resistance, information which is then used at Block (F) to determine the required plastic modulus of the beam section. If the largest available plastic modulus is too small to satisfy the current problem then there can be no solution. This eventuality is checked at Block (G). If the section list is found to be too short then a message to this effect is output at Block (J); otherwise Block (H) is entered and the required plastic modulus is compared with those available. The first section in the list having a plastic modulus in excess of that required is adopted and the result is output at Block (I).

Since the conclusion to a specific problem is known when either Block (I) or (J) is entered the program run could be terminated there. However, it is probable that in a real design situation a number of beams would need to be processed. For this reason the routes from Blocks (I) and (J) both return to the input of new problem data at Block (B). And in order not to trap the designer in a closed loop an exit from the program operates at Block (C) when the span is input as zero.

1.3.4 Description of Program PD1

The program is listed below. Of the forty-five statements needed to write this program to the specification given in Section 1.3.1 only six of them (Statements 310–360) are directly related to design. The remainder are concerned with the input and output of information. In the following description program block reference letters are the same as those given in the flow diagram shown in Fig. 1.2. A list of the variables and arrays used in Program PD1 is given below:

A1	Distance of load from left hand support
F1	Load factor
I	Counter of array rows
J	Counter of array columns
L1	Beam span
M1	Required plastic moment of resistance
S( )	Section property array
S1	Required plastic modulus
W1	Load
X	Left hand support condition
Y	Right hand support condition
Y1	Steel yield stress

```
100 FOR I=1 TO 4
110 FOR J=1 TO 4
120 READ S(I,J)
130 NEXT J
140 NEXT I
150 DATA 4845E+2,162E+4,5512E+3,1093E+4
160 DATA 254,457,610,914
170 DATA 146,152,305,305
180 DATA 37,74,179,253
190 PRINT "SPAN (M), LOAD (KN) AND DISTANCE OF LOAD"
200 PRINT "FROM LEFT HAND SUPPORT (M) =";
210 INPUT L1,W1,A1
220 IF L1=0 THEN 540
230 PRINT
240 PRINT "L.H. SUPPORT CONDITION - IF FIXED TYPE 1 ELSE 0";
250 INPUT X
260 PRINT "R.H. SUPPORT CONDITION - IF FIXED TYPE 1 ELSE 0";
270 INPUT Y
280 PRINT
290 PRINT "LOAD FACTOR AND STEEL YIELD STRESS (N/MM↑2) =";
```

```
300 INPUT F1,Y1
310 LET M1=F1*W1*A1/(X+L1/(L1-A1)+Y*A1/(L1-A1))
320 LET S1=M1*10↑6/Y1
330 IF S(1,4)<S1 THEN 460
340 FOR J=1 TO 4
350 IF S(1,J)>S1 THEN 370
360 NEXT J
370 PRINT
380 PRINT "CALCULATED PLASTIC MODULUS ="S1"MM↑3"
390 PRINT "    CHOSEN PLASTIC MODULUS ="S(1,J)"MM↑3"
400 PRINT "SECTION DIMENSIONS ARE :"
410 PRINT "DEPTH="S(2,J)"MM","WIDTH="S(3,J)"MM"
411 PRINT "UNIT WEIGHT OF SECTION="S(4,J)"KG/M"
420 PRINT
440 PRINT
450 GOTO 190
460 PRINT
470 PRINT "THE LARGEST AVAILABLE SECTION IS TOO SMALL"
480 PRINT "REQUIRED PLASTIC MODULUS ="S1"MM↑3"
490 PRINT "MAXIMUM AVAILABLE PLASTIC MODULUS ="S(1,4)"MM↑3"
500 PRINT
510 PRINT
520 PRINT
530 GOTO 190
540 END
```

1.3.4.1 Block (A): Statements 100–180

The permanent design data are read into an array called S( ) which has four rows, each containing four elements. The first row of this array holds values of plastic modulus in order of ascending section weight. The second, third and fourth rows store the depth, breadth and unit weight of each section respectively. Thus each column of elements contains the numerical information relating to one section.

The statements at lines 100 to 140 cause this information to be read into the array. Because the READ statement at line 120 is obeyed 16 times (see the statements at lines 100 and 110) the computer looks for 16 numbers. These are provided by the four numbers in each of the four DATA statements at lines 150 to 180. If all the data could have been contained in a single line then this too would have been acceptable; it is only the order in which the information is presented that is important. Generally it is simpler to present this information row by row (as in this program) for ease of checking. And for this reason too it is advisable, though not essential, to relate a datablock to its own array by placing them together in the program.

1.3.4.2 Blocks (B), (C) and (D): Statements 190–300

Program Blocks (B) and (D) comprise the second part of the data input. They have been programmed with statements of the type discussed in Section 1.2.9.3, where it was suggested that an INPUT statement should be preceded by a PRINT statement which gives meaning to the question mark that automatically signals any demand for input. In this example the statements at lines 190 and 200 not only remind the designer that the computer needs values for the span of the beam, the magnitude of the load and its position in the span, but also the units in which these quantities must be presented.

The single statement at line 220 comprises Block (C). It is a conditional statement which causes a jump (see Section 1.2.8.1) to the end of the program (and hence an end to the program run) if the input span value is zero.

The statements at lines 240 to 260 show how a multiple choice of input information can be effected. In this problem a support condition may take one of two forms, simple or fixed. The values 0 or 1 assigned by the designer to the variables X and Y will later influence the calculated value of the plastic moment of resistance by affecting those terms in equation (1.1) which represent the internal work done by the support hinges.

1.3.4.3 Block (E): Statement 310

This single LET statement comprises the structural analysis, which in this case is simply a reiteration of equation (1.2) in BASIC language. The statement says that M1, the required plastic moment of resistance, is equal to the result of carrying out the calculation on the right hand side of the expression.

The programmed derivation of M1 could equally have followed the formal steps of the manual analysis. Instead of the present statement at line 310 the following could have been substituted:

Referring to equation (1.1), and assuming unit hinge rotation at support B, then if E1 is the external work done by the moving load,

305 LET $E1 = F1*W1*A1$

And if I1 is the internal work done by the plastic hinges in rotating then:

306 LET $I1 = X*L1/(L1 - A1) + Y*A1/(L1 - A1)$

The previous statement at line 310 would then be written as:

310 LET $M1 = E1/I1$

which is a restatement of equation (1.2). Thus either the final form of an algebraic expression or the steps which lead up to it may be programmed to suit the convenience of the programmer and the possible need to output intermediate information.

1.3.4.4 Block (F): Statement 320

This LET statement, which represents equation (1.3), calculates the required plastic modulus, S1.

1.3.4.5 Block (G): Statement 330

Before attempting to adopt a section this statement verifies whether there is, in fact, one available. It does this by comparing the largest plastic modulus value included in the list, S(1,4), with that required. If S(1, 4) is greater than S1 then Block (G) is entered; otherwise there is no solution and a jump to Block (J) causes a 'fail' message to be output.

1.3.4.6 Block (H): Statements 340–360

The purpose of this block of statements is to find and adopt the lightest structural steelwork section in the stored list which satisfies the design criterion. It does this by inspecting the plastic modulus values from the first row of array S( ) in turn until one is found which exceeds the required value, S1. The location (J) of this section in the first row then defines the adopted section and its properties.

In programming terms this procedure requires three statements. The statement at line 340 defines the value of J, and hence the column elements in the array, at which comparisons will be made. The content of an element E(1, J) is compared with S1 at line 350. If it is smaller than S1 then the statement at line 360 advances the value of J and the process is repeated. If it is larger than S1 then the conditional jump at line 350 halts the process and the latest value of J is preserved in the variable called J, thus defining the adopted section.

1.3.4.7 Blocks (I) and (J): Statements 370–530

These blocks of program are concerned with the output of information, either the solution to the problem in the case of Block (I) (see lines 370 to 450) or a reason for failure to produce a solution in the case of Block (J) (see lines 460 to 530). Block (I) is entered directly from Block (H); Block (J) is entered as the result of a jump from Block (G). When the computer reaches either of the lines 450 or 530 the program user is returned to a request for new problem data at line 190.

The PRINT statements in these two blocks either output newlines (e.g. the statement 420 PRINT) to give vertical spaces between results, or they are a mixture of text and variable output of the kind discussed in Section 1.2.10.

1.3.4.8 Block (K): Statement 540

This statement defines the end of the program.

1.3.4.9 Example of Program PD1 Output

Typical results given by this program appear below. Note that in this output a question mark signifies a stage at which the computer halted to receive an input of data and that the numbers following a question mark match the number of items requested by the preceding text. On receiving the last item of data the computer responds with a solution.

The propped cantilever in the first example has a span of 8.4 m and carries a point working load of 176 kN situated 3.2 m from the left hand (fixed) support; the load factor in this case is 1.75 and the steel yield stress is 250 N/mm². To satisfy these parameters the second section in the list was adopted. In the second example, a simply supported beam spanning 10 m and carrying a centrally situated point load of 760 kN, the largest section in the list was found to be too small.

```
PD1

SPAN (M), LOAD (KN) AND DISTANCE OF LOAD
FROM LEFT HAND SUPPORT (M) = ? 8.4,176,3.2

L.H. SUPPORT CONDITION - IF FIXED TYPE 1 ELSE 0 ? 1
R.H. SUPPORT CONDITION - IF FIXED TYPE 1 ELSE 0 ? 0

LOAD FACTOR AND STEEL YIELD STRESS (N/MM†2) = ? 1.75,250

CALCULATED PLASTIC MODULUS = 1.50739E+6 MM†3
   CHOSEN PLASTIC MODULUS = 1.62E+6 MM†3
SECTION DIMENSIONS ARE :
DEPTH= 457 MM  WIDTH= 152 MM
UNIT WEIGHT O F SECTION= 74 KG/M


SPAN (M), LOAD (KN) AND DISTANCE OF LOAD
FROM LEFT HAND SUPPORT (M) = ? 10,760,5

L.H. SUPPORT CONDITION - IF FIXED TYPE 1 ELSE 0 ? 0
R.H. SUPPORT CONDITION - IF FIXED TYPE 1 ELSE 0 ? 0

LOAD FACTOR AND STEEL YIELD STRESS (N/MM†2) = ? 1.75,250

THE LARGEST AVAILABLE SECTION IS TOO SMALL
REQUIRED PLASTIC MODULUS = 1.33E+7 MM†3
MAXIMUM AVAILABLE PLASTIC MODULUS = 1.093E+7 MM†3


SPAN (M), LOAD (KN) AND DISTANCE OF LOAD
FROM LEFT HAND SUPPORT (M) = ? 0,0,0


RUNNING TIME:   1.5 SECS   I/O TIME :   4.9 SECS
```

1.4 Automatic Design v. Decision Design—Two Routes to a Solution

There are two fundamentally different approaches to the programming of structural design problems. One method, which will be called *Automatic Design*, is to program the solution to the problem in such a way that when the computer receives information about the structural geometry, loading and material properties a solution is automatically generated, usually as the result of an iterative process. The computer is programmed to attempt a succession of designs, each of which is based upon updated information elicited from the previous design. A solution is achieved when two consecutive designs produce identical results. From the designer this type of program requires only that he chooses realistic values for his basic design parameters. He cannot influence the progress of a solution since the design techniques are an inherent part of the program.

The example given in Section 1.3 was a simple illustration of the automatic design process; but in that case, because of the nature of the problem, only one pass through the analysis and design phases was necessary to achieve a solution.

In contrast with this, the second method, called *Decision Design*, demands the active cooperation of the designer in obtaining a solution. By this method the problem is programmed in such a way that at appropriate stages in the calculation the computer halts and calls for information. These stages coincide with those at which a manual designer would make a decision—possibly to increase a section size, decide upon a cable eccentricity or the magnitude of moment redistribution. The calculation is only able to proceed when the designer has input his decision. The attraction of this method is that the designer retains as much control over the course of his solution as he would in an equivalent manual design attempt.

A decision design approach to the example given in Section 1.3 would have been to program for the computer's acceptance or rejection of sections suggested by the program user.

Of these two methods the programming of an automatic design is the more complex because all foreseeable circumstances must be provided for. Even though the use of this method largely sacrifices the intelligent anticipation that a manual designer brings to his solution in favour of an inefficient (albeit successful) iterative procedure, at least some design experience must be programmed. But because the path to some design decisions is often obscure a substantial part of the programming effort is given to examining the nature of decisions and devising suitable programmed equivalents.

The costs of computer aided design are affected by the kinds of programs which are developed and used. And whilst the least expensive overall approach is not necessarily the best it should at least be considered so that comparisons may be made. Computer aided design costs arise from the three following sources:

1. Programming effort—i.e. the time that it takes to write and prove a program. For the reasons already outlined this is greater in the case of an automatic design based program than for its decision design equivalent. Assuming that the program may be used to solve a range of problems then the development costs will be written off over a number of jobs.
2. Computing time—i.e. the time that a computer is actually engaged in executing a program. The cost of this item is a function of the nature of the program, the efficiency with which it performs its task and the unit cost of running the computer system. Given comparable automatic design and decision design programs the former will often require a significantly greater amount of computing time than the latter. This is because an automatic design usually follows an iterative path which must be pursued to its

inevitable conclusion. In contrast with this, even though a decision design solution may also be iterative the program user can often anticipate the solution and thereby speed its convergence.

3. Designer time—i.e. the time that a designer is engaged in preparing and handling design data, running the program and evaluating results. This cost will be substantially less in the case of an automatic design program run and the difference will become even more significant as the complexity of the problem to be solved increases.

Whilst decision design techniques are valid in all design situations they are most appropriate when applied to the solution of 'open ended' problems, ones for which a given set of design parameters leads to an infinity of solutions. This class of problem includes designs in prestressed concrete, general reinforced concrete and plated steelwork, or indeed any problem for which the options must be left open as long as possible. The quality of a decision design result is contingent on the soundness of the program user's decisions.

Automatic design techniques are only applicable to 'closed' problems, ones for which a given set of parameters leads to a unique solution, e.g. designs which are based upon a choice of standard sections, or any other problem for which a sufficient number of constraints can be introduced. In this case the quality of the solution rests with the program.

An automatic design program and its data form a package which requires no external directive for it to produce a result once it is in the computer. Such a program may either be batch processed or run on an on-line operation.

Batch processing is a means whereby the program and its data, prepared on punched cards or paper tape, are processed at the site of the computer, usually by a computer operator. In this case the designer will only have been in communication with the machine at second hand. In contrast with this, an on-line operation allows the designer to be in direct contact with the computer. Either type of program may be run in this way but it is particularly apt in the case of decision design programs. The computer is contacted through the medium of a remotely situated teleprinter terminal. Instructions and data may be input to the computer on punched paper tape although it is more usual to convey them directly to the machine via the teleprinter keyboard. This makes it a simple matter for the designer to run programs and influence their course.

1.5 Communicating with the Machine

1.5.1 A Control Language

A *Control Language* comprises a series of English words and teleprinter keyboard responses. The language allows a simple dialogue to take place between a user and the machine thus making it possible for the machine to interrogate a user about his intentions and for a user to instruct the machine in familiar terms on which tasks it has to perform. When the computer is contacted, usually by telephone, it responds by causing the teleprinter to type key words to which the user will reply through his keyboard in the explicit way laid down by the control language. Each computer system is addressed in a particular way and the rules for this are described in the relevant user manual. A brief description of the Series 400 Time Sharing Programming System control language follows to aquaint the reader with a typical man—machine dialogue situation. For full information about this particular system the reader should see Ref. 2.

A specific region of the computer's long-term memory is set aside to store the programs and data created by an individual user. For reasons of security it is therefore essential that he alone gains access to that part of the machine; to this end he is allotted a User Identification Number and a Password.

When contact with the machine is established at the beginning of a computing session the teleprinter types USER ID—, to which the user replies by typing his number, 2935 say, and depressing the RETURN key. (No user response or command is understood or executed by the computer until that key has been pressed.) If the computer recognizes this as being a valid User Number then the teleprinter types PASSWORD, followed on the next line by five overprinted characters and a question mark, thus ᴇᴇᴇᴇᴇ?. The carriage then returns to the beginning of that line. The user's response is to type his password onto the overprinted characters and in this way the integrity of the password is preserved.

If the password is recognized by the computer as relating to the identification number it has already accepted then the teleprinter types TYPE OLD OR NEW:. The object of this request is to ascertain the nature of the operation to be carried out. If the user replies OLD then this warns the computer that its task will be to edit or run an existing program which is already held in its memory (the permanent file). A response of NEW would inform the machine that a new program was to be created. Either response causes the computer to request a program title; this is signified by the teleprinter typing FILE NAME:.

If the user replies with the program title, RC5 say, then the effect that this has depends upon his previous response. If it was OLD then a copy of an existing program called RC5 would be made in the temporary file in readiness for some activity to happen; in this event the program in its original form would remain in the permanent file. If the previous response was NEW then RC5 becomes the name of a new program which will be created on the temporary file, but for which at this time there is no permanent record.

When the computer is informed of a program title then the teleprinter types READY and the system waits for a further response from the user. On the teleprinter paper a typical opening dialogue appears as:

```
USER  ID—2935
PASSWORD
ᴇᴇᴇᴇᴇ?
```

TYPE OLD OR NEW: NEW
FILE NAME:RC5
READY

A new program is usually entered into the computer line by line via the teleprinter keyboard. The order in which individual lines are entered is unimportant because they will automatically be stored according to their line numbers. To signify that the program is complete the highest numbered line will always be an END statement.

Any information which is currently registered in the temporary file (a copy of an existing program or a newly created one) will be lost if either another program is called or the user signs himself out of the system. The contents of the temporary file can however be stored permanently in the computer's memory by typing the command SAVE before another activity is embarked upon.

1.5.2 Tracing Program Errors

When the programmer types the command RUN the computer will respond in one of four different ways:
1. Due to language errors (in syntax, mis-spelling of key words, incorrect use of brackets, etc.) the program may not work at all. In this case the computer activates the teleprinter to type out each line number at which an error occurs, together with a brief diagnosis of the nature of the error.
2. The program may execute calculations but produce incorrect results. This would suggest that there are errors in the logic of the program or in the form of one or more of the mathematical expressions.
3. A variant of (2) is for the program to work but to produce no results at all, either because for some reason the output statements have been circumvented or the calculation has entered a loop from which no exit has been provided.
4. The program may behave perfectly in all respects.

In the unlikely event of case (4) happening there is no more to be said. The other situations all require that errors be recognized and corrected.

Case (1) errors are the simplest to rectify because they have already been identified by the computer. They are corrected by retyping the whole statement (including the line number) in its true form and re-entering it into the computer. This action effectively overwrites the previously incorrect statement with that number.

Case (2) and (3) programming errors are seldom found by inspecting the listed program in its original form. They are most easily identified by inserting extra PRINT statements into the program to aid diagnosis. These can take one of two forms depending upon whether the arithmetic of the calculation or the logic of its route is being checked. If the former, then it is often helpful to output the results of intermediate steps in the calculation. In the latter case the output of identifiable text at strategic points in the program often helps to check the logic of loops and jumps. When errors have been located the incorrect statements can be overwritten as before.

The wisdom of leaving gaps between the line numbers of consecutive statements should now be obvious. Unwanted diagnostic statements can be eliminated by using the EDIT DELETE facility.

1.5.3 Testing the Program

Once the programmer is reasonably certain that all the errors have been rectified then the program should be thoroughly tested. This is accomplished by setting the program a whole range of problems assumed to be within its competence. In this respect it is important to check upon its behaviour when operating close to its upper and lower bounds as well as within the normal middle range. If it is found that some combinations of circumstances cannot be handled by the program, and it is decided that the extra programming effort needed to rectify the situation cannot be justified, then these limitations should be clearly stated in the program documentation.

References

1. Series 400 BASIC LANGUAGE for TIME SHARING PROGRAMMING SYSTEM, Reference Manual BR 03, Honeywell Information Systems Ltd.
2. Series 400 TERMINAL USER'S MANUAL for TIME SHARING PROGRAMMING SYSTEM, Reference Manual BR 04, Honeywell Information Systems Ltd.

Chapter 2

Automatic Design and Decision Design Methods Applied to Reinforced Concrete Slabs

2.1 Introduction

Because slabs are designed as strips of unit width the programmer is not concerned with establishing a relationship between the depth and breadth of a section. Moreover, the singly reinforced section depth found to be structurally necessary·is usually acceptable. These two factors taken together make slab design a 'closed' problem which yields a unique solution for each set of design parameters. Given this situation both automatic design and decision design methods are equally applicable and the choice of method is largely a matter of preference. Both are discussed in detail in Sections 2.2. and 2.3.

The scope of both these programs was purposely limited to that of designing simply supported slabs in order to focus attention on matters leading to the acceptance of a reinforced concrete section; but considering the programming effort involved such a limitation would be unacceptable in practice. Section 2.4 looks at the slab design problem in greater depth. There it is shown how to program one method of analysing two-way spanning slabs and how to incorporate these results with the decision design based program RC9.

2.2 The Automatic Design of Reinforced Concrete Slabs

2.2.1 Program Specification—RC4

The purpose of this program is to determine the depth and the reinforcement required by simply supported, singly reinforced slabs of up to 20 m span, for all exposure conditions. The slabs are designed for the ultimate limit state of bending and shear and the serviceability limit states of deflection and cracking as defined in CP110: Part 1:1972. If shear reinforcement is necessary then the amount required at the support is given.

2.2.2 The RC4 Flow Diagram

This flow diagram is shown in Fig. 2.1. For a given set of design parameters a slab solution will be considered acceptable when the individual criteria for bending, shear, deflection and cracking have been satisfied. The first design assumption made in the program is that a 'balanced' design is a possible solution. This will either be proved or disproved at a later stage in the design process. The calculation is initially forced onto the 'balanced' design route by setting a *switch*, called T8 in the program, to zero. Whilst various other criteria are tested, T8 will remain zero if they are all satisfied, or will take the value of 1, 2 or 3 as and when another design criterion becomes dominant.

Referring to Fig. 2.1 and tracing the most direct route to a solution, it will be seen that a 'balanced' design is achieved in the following way. Ignoring Program Blocks (A), (B), and (C), which are common to all problems, the 'balanced' design route is:

$$(T8 = 0) \to D \to E \to F \to G \to H \to I \to L \to M \to N \to O \to P \to Q$$

On the way through this route a loop is initiated at Block (I) to determine the final 'balanced' design dimensions. In common with designs based upon the minimum depth and deflection criteria Block (P) will be circumvented if shear reinforcement is not required. Otherwise the route through to a 'balanced' design is a direct one.

It may happen that the 'balanced' design depth is less than that chosen by the designer as being the minimum practical construction depth. In this case T8 is set equal to 1 at Block (H) and the calculation enters an 'under-reinforced' routine at Block (J). The route followed by the minimum practical depth criterion is:

$$(T8 = 1) \to D \to E \to F \to G \to H \to J \to E \to F \to G \to K \to L \to M \to N \to O \to P \to Q$$

When T8 = 1 the 'under-reinforced' routine loop ensures that the correct effective depth and area of reinforcement have been found before the calculation re-enters the main body of the program. Assuming that the deflection and shear criteria are satisfied the results at Block (Q) will be those for a slab of minimum practical depth.

For both 'balanced' and minimum depth slabs their effective depths are checked against the deflection requirements at Block (L). If they do not meet this criterion then T8 is set equal to 3 and the calculation enters a loop:

$$(T8 = 3) \to L \to R \to E \to F \to G \to S \to M$$

which ensures that at this stage a section is derived which meets the requirements of both bending and deflection.

For the design criteria so far considered unique solutions can be found. The design for shear is less straightforward. In this respect the designer himself must decide what is, and what is not, acceptable. Within the 'normal' bounds of span and loading shear is not usually a problem to be reckoned with in slab design and in most cases a simple check on the level of shear stress is all that is required. If the problem is to be investigated further then the following approach is suggested.

At Block (M), for the largest section so far derived, the actual and allowable shear stresses are calculated. Three courses may then be pursued.

1. If the actual shear stress is less than the allowable value then the route becomes:

$$(T8 = 0, 1 \text{ or } 3) \to M \to N \to O \to Q$$

2. If the actual shear stress is greater than the allowable value but less than the maximum permitted shear stress then the current maximum slab depth is preserved and vertical shear reinforcement is designed on that basis. This procedure is defined by the route:

$$(T8 = 0, 1 \text{ or } 3) \to M \to N \to O \to P \to Q$$

3. If the actual shear stress is greater than the maximum permitted shear stress then the depth of the section is increased to reduce the level of shear stress to the maximum permitted value. Shear reinforcement is then provided for that section. In this case T8 becomes equal to 2 and the solution follows the route:

$$(T8 = 2) \to M \to N \to T \to J \to E \to F \to G \to U \to M \to P \to Q$$

2.2.3 Writing Program RC4

The approach adopted in writing this program followed the general lines laid down in Section 1.3.3.1. The heart of any reinforced concrete design program lies in the translation of a calculated steel area into a number of bars of given diameter arranged in a specific way. Once this information is known the distance between the centre of gravity of the reinforcement and the tension face may be calculated and hence the effective depth of the section follows. For these reasons Blocks (E), (F) and (G) were developed first. As general progress was made Blocks (A), (B) and (Q) were expanded to handle the input and output of increasing amounts of information. The remainder of the program was developed in the following six stages:

1. Blocks (D) and (I) Balanced Design
2. Blocks (H), (J) and (K) Minimum Depth Design
3. Blocks (L), (R) and (S) Design for Deflection

4. Block (M)   Shear Checks
5. Blocks (N), (O), (T) and (U)   Shear Depth Design
6. Block (P)   Shear Reinforcement

It will be appreciated that whilst program development generally followed these lines the actual process was not as straightforward as has been implied; inevitably, later stages in development affected earlier stages to some degree. Progress was assisted by developing both the flow diagram and program in parallel. Program development was further aided by maintaining a constantly updated list of the variables and arrays used.

2.2.4 Description of Program RC4

A list of the variables and arrays used in programs RC4 and RC9 (for a description of RC9, see Section 2.3) is given below. Those identified by a single asterisk appear only in Program RC4 whilst those having a double asterisk belong to Program RC9. Those common to both programs have no asterisk.

A( )	Contains the table of bar areas
A1	Steel area or area/row
*B( )	Contains the calculated spacing for vertical shear reinforcements
*B1	The sum of bar diameter and spacing
*B2	The difference between consecutive solutions
*B3	The difference between alternate solutions
*B7	Rounded down value of shear reinforcement spacing
*B8	Lever arm factor
*B9	Area of single shear reinforcing bar
C2	Cover chosen from array F( )
D1	Effective depth
D2	Distance from tension face to c.g. of reinforcement
D5	Effective depth required for deflection criterion
*E	Array column counter
*F	Array row counter
F1	Characteristic steel strength
F2	Characteristic concrete strength
F( )	Contains the contents of Table 19 CP 110 (nominal cover to reinforcement)
H1	Overall slab depth before rounding off
H3	Overall slab depth after rounding off
I1	Locates row of bar area table containing chosen steel area
I	Array row counter
J1	Locates column of bar area table containing chosen steel area
J	Array column counter
L1	Span
M( )	Contains selected coefficients from Table 10 CP 110 (Modification factors for tension reinforcement)

M1	Ultimate bending moment
*M2	Steel moment of resistance for deflection criterion
*M3	Applied bending moment
*M8	Takes the greatest of the values of M2 and M9
M9	Concrete moment of resistance for deflection criterion
N7	Sets a limit to J in array A( )
N8	Theoretical maximum bar spacing
*O8	Switch
*O9	Exposure condition
P1	Steel percentage at section of maximum bending moment
P2	Steel percentage at support
P5	Individual modification factor selected from array M( )
P9	Largest preferred bar diameter
P( )	Contains the preferred bar diameters
Q1 Q2 Q3 Q4	Intermediate steps introduced to simplify interpolation expressions
**Q7	Switch
**Q8	Switch
*R1 *R2	Hold steel areas for comparison
R3	Number of rows of reinforcement
R9	Shear stress factor adopted from array R( )
R( )	Contains the contents of Table 14 CP 110 (Factors which increase the allowable shear stress in shallow slabs)
*S	The sum of the 3rd, 4th and 5th rows in array S( )
S1 S2 S3 S4	Hold information used in bar spacing calculation
S( )	Contains part of Table 24 CP 110 (For zero redistribution of moment, the allowable clear distance between bars related to characteristic steel strength and slab depth
*T	Sets steel percentage for deflection criterion design
*T8	Defines design criterion
T9	Minimum acceptable slab depth
U	Array row counter
V	Array column counter
V4	Actual shear stress
V8	Absolute maximum shear stress
V9	Allowable shear stress
V( )	Contains Tables 5 and 6 CP 110 (ultimate shear stress in slabs and maximum value of shear stress in slabs)
W1	Characteristic superimposed load

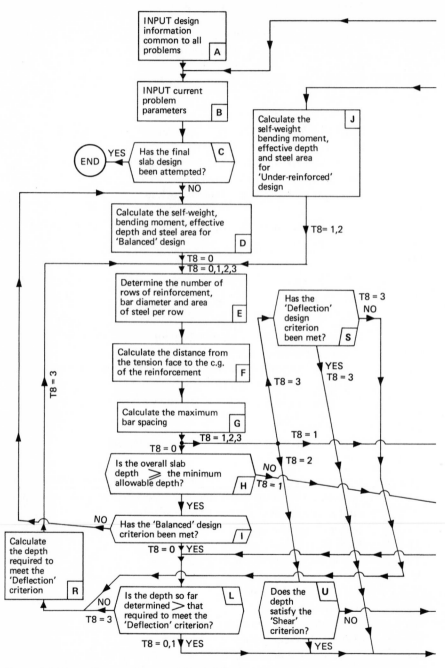

Figure 2.1 Flow diagram for program RC4 (the automatic design of reinforced concrete slabs)

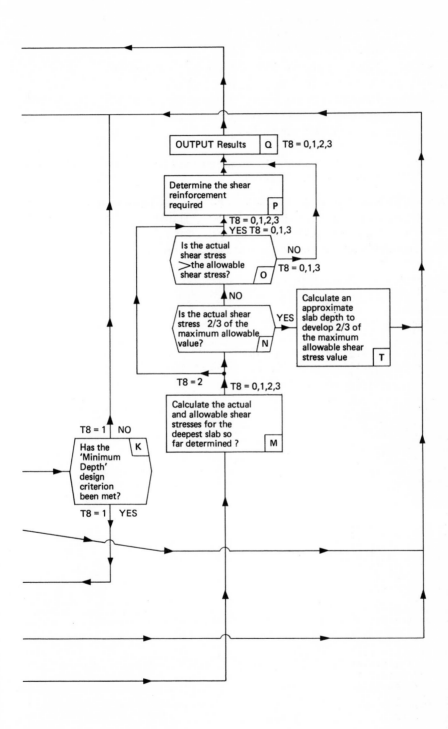

OUTPUT Results | Q | T8 = 0,1,2,3

Determine the shear
reinforcement
required | P

T8 = 0,1,2,3
YES T8 = 0,1,3

Is the actual
shear stress
> the allowable
shear stress? | O | NO
T8 = 0,1,3

NO

Is the actual shear
stress 2/3 of the
maximum allowable
value? | N | YES

Calculate an
approximate
slab depth to
develop 2/3 of
the maximum
allowable shear
stress value | T

T8 = 2 | T8 = 0,1,2,3

Calculate the actual
and allowable shear
stresses for the
deepest slab so
far determined ? | M

T8 = 1 | NO

Has the
'Minimum
Depth'
design
criterion
been met? | K

T8 = 1 | YES

*W2	Current dead load
W3	Unit weight of concrete
*W4	New dead load
*W( )	Holds successive slab weights for comparison
*X	Array row counter
*X1	Iteration counter
*X( )	Holds slab solutions for reference
Y	Array column counter

Z1)
Z2 } Elements of quadratic equation used to determine steel area in
Z3) under-reinforced section

Program RC4 is listed below and should be read in conjunction with the flow diagram shown in Fig. 2.1. Individual blocks of statements play an identifiable role in the design process and form the basis of the discussion which follows. The statement blocks in this program are labelled (A) to (U).

```
100 DIM A(10,23)
110 FOR J=2 TO 23
120 A(1,J)=75+(J-2)*25
130 NEXT J
140 FOR I=2 TO 10
150 READ A(I,1)
160 NEXT I
170 DATA 6,8,10,12,16,20,25,32,40
180 FOR I=2 TO 10
190 FOR J=2 TO 23
200 A(I,J)=(250*3.142*A(I,1)*2)/A(1,J)
210 NEXT J
220 NEXT I
230 FOR J=1 TO 7
240 READ B(1,J)
250 NEXT J
260 DATA 6,8,10,12,16,20,25
270 FOR X=1 TO 5
280 FOR Y=1 TO 2
290 READ R(X,Y)
300 NEXT Y
310 NEXT X
320 DATA 250,1.0
330 DATA 225,1.05
340 DATA 200,1.1
350 DATA 175,1.15
360 DATA 150,1.2
370 FOR X=1 TO 5
380 FOR Y=1 TO 5
390 READ F(X,Y)
400 NEXT Y
410 NEXT X
420 DATA 20,25,30,40,50
430 DATA 25,20,15,15,15
440 DATA 0,40,30,25,20
450 DATA 0,50,40,30,25
460 DATA 0,0,0,60,50
470 FOR X=1 TO 7
480 FOR Y=1 TO 5
490 READ V(X,Y)
500 NEXT Y
510 NEXT X
```

```
520 DATA 0,20,25,30,40
530 DATA 0.25,0.35,0.35,0.35,0.35
540 DATA 0.50,0.45,0.5,0.55,0.55
550 DATA 1.0,0.6,0.65,0.7,0.75
560 DATA 2.0,0.8,0.85,0.9,0.95
570 DATA 3.0,0.85,0.9,0.95,1.0
580 DATA 0,2,24,2.51,2.75,3.18
590 FOR X=1 TO 3
600 FOR Y=1 TO 5
610 READ S(X,Y)
620 NEXT Y
630 NEXT X
640 DATA 250,410,425,460,500,
650 DATA 300,185,180,165,150
660 DATA 250,250,250,200,200
670 FOR U=1 TO 6
680 FOR V=1 TO 9
690 READ M(U,V)
700 NEXT V
710 NEXT U
720 DATA 0,0.25,0.5,0.75,1.0,1.5,2.0,2.5,3.0
730 DATA 250,2.0,1.98,1.62,1.44,1.24,1.13,1,06,1.01
740 DATA 410,1.6,1.23,1.09,1.0,0.9,0.84,0.8,0.77
750 DATA 425,1.55,1.2,1.06,0.98,0.88,0.83,0.79,0.76
760 DATA 460,1.41,1.11,0.99,0.92,0.84,0.78,0.75,0.72
770 DATA 500,1.27,1.03,0.92,0.86,0.79,0.74,0.71,0.68
780 DIM W(50)
790 N7=23
800 PRINT "CONCRETE GRADE (FCU) TO BE CHOSEN FROM THE FOLLOWING"
810 PRINT "20,25,30,40 AND 50 N/MM+2"
820 PRINT
830 PRINT "STEEL GRADE (FY) TO BE CHOSEN FROM THE FOLLOWING"
840 PRINT "250,410,425,460 AND 500 N/MM+2"
850 PRINT
860 GO TO 950
870 PRINT "NEXT PROBLEM"
880 FOR I=1 TO 4
890 FOR J=1 TO 10
900 X(I,J)=0
910 NEXT J
920 NEXT I
930 PRINT
940 PRINT
950 PRINT "SPAN(M),SUPERLOAD(KN/M+2),CONCRETE WEIGHT(KN/M+3),"
960 PRINT
970 PRINT "MINIMUM SLAB DEPTH(MM),FY(N/MM+2),FCU(N/MM+2),"
980 PRINT
990 INPUT L1,W1,W3,T9,F1,F2
1000 PRINT
1010 IF L1<=20 THEN 1050
1020 PRINT "SPAN GREATER THAN 20M."
1030 PRINT
1040 GO TO 950
1050 IF L1=0 THEN 4740
1060 PRINT
1070 PRINT "TYPE 1,2,3 OR 4 FOR MILD,MODERATE,SEVERE OR VERY SEVERE"
1080 PRINT
1090 PRINT "EXPOSURE CONDITIONS";
1100 INPUT 09
1110 PRINT
1120 FOR Y=1 TO 5
1130 IF F(1 Y)=F2 THEN 1150
1140 NEXT Y
1150 C2=F(09+1,Y)
1160 IF C2>0 THEN 1310
1170 FOR Y=1 TO 5
```

```
1180 IF F(O9+1,Y)>0 THEN 1200
1190 NEXT Y
1200 C2=F(O9+1,Y)
1210 F2=F(1 Y)
1220 PRINT "MINIMUM FCU FOR THESE EXPOSURE CONDITIONS="F(1,Y)"N/MM*2"
1230 PRINT
1240 PRINT "IF THIS VALUE OF FCU IS ACCEPTABLE TYPE 1 ELSE 0";
1250 INPUT O8
1260 PRINT
1270 IF O8=1 THEN 1310
1280 PRINT "YOU MUST REASSESS EITHER FCU OR EXPOSURE CONDITIONS"
1290 PRINT
1300 GO TO 950
1310 PRINT "AVAILABLE BAR DIAMETERS ARE 6,8,10,12,16,20,25,32,40MM"
1320 PRINT
1330 PRINT "PREFERRED DIAMETERS";
1340 INPUT P(1),P(2),P(3),P(4),P(5),P(6),P(7),P(8),P(9)
1350 PRINT
1360 T8=0
1370 X1=0
1380 W2=T9*W3/1000
1390 M1=(1.6*W1*L1+2)/8+(1.4*W2*L1+2)/8
1400 D1=SQR((M1*10+3)/(0.15*F2))
1410 A1=10+6*M1/(0.6525*F1*D1)
1420 GO TO 1610
1430 T8=1
1440 X1=0
1450 GO TO 1500
1460 T8=2
1470 X1=0
1480 GO TO 1500
1490 T9=L1*(1.6*W1+1.4*H3*W3/1000)/(4*V8/3)+D2
1500 M1=(1.6*W1*L1+2)/8+(1.4*T9*W3*L1+2)/8000
1510 D1=T9-D2
1520 Z1=F1*D1/1.15
1530 Z2=3780.718*F1*F1*M1/F2
1540 Z3=F1*F1/(529*F2)
1550 A1=(Z1-SQR(Z1+2-Z2))/Z3
1560 B8=1-(A1*F1)/(920*F2*D1)
1570 IF B8<0.95 THEN 1590
1580 A1=M1*10+6/(F1*0.95*D1)
1590 IF A1>=2.5*D1 THEN 1610
1600 A1=2.5*D1
1610 R3=1
1620 IF T8<3 THEN 1640
1630 A1=P1*D5*10
1640 FOR X=1 TO 9
1650 IF P(X)=0 THEN 1670
1660 P9=X
1670 NEXT X
1680 IF A1>A(P9+1,2) THEN 1700
1690 GO TO 1760
1700 R3=R3+1
1710 A1=A1*(R3-1)/R3
1720 GO TO 1680
1730 I1=P9+1
1740 J1=2
1750 GO TO 1880
1760 R2=1000000
1770 FOR I=2 TO 10
1780 FOR J=2 TO N7
1790 IF P(I-1)=0 THEN 1870
1800 R1 =A(I,J)-A1
1810 IF R1<0 THEN 1860
1820 IF R1>R2 THEN 1860
1830 LET R2=R1
```

```
1840 LET I1=I
1850 J1=J
1860 NEXT J
1870 NEXT I
1880 IF C2>A(I1,1) THEN 1900
1890 C2=A(I1,1)
1900 D2=(20*(R3-1)/2)+(R3*A(I1,1)/2)+C2
1910 IF T8<3 THEN 1930
1920 D1=D5
1930 P1=100*A1*R3/(1000*D1)
1940 FOR Y=1 TO 5
1950 IF S(1,Y)=F1 THEN 1970
1960 NEXT Y
1970 S1=S(2,Y)
1980 IF P1>1.0 THEN 2040
1990 IF P1<0.5 THEN 2020
2000 S2=S1/P1
2010 GO TO 2050
2020 S2=S1*2
2030 GO TO 2050
2040 S2=S1
2050 S3=1000000
2060 IF (D1+D2)<=S(3,Y) THEN 2080
2070 GO TO 2090
2080 S3=3*D1
2090 IF S2>S3 THEN 2120
2100 N8=S2
2110 GO TO 2140
2120 N8=S3
2130 N8=N8+A(I1,1)
2140 FOR J=2 TO 23
2150 IF (N8-A(1,J))<0 THEN 2170
2160 NEXT J
2170 N7=J-1
2180 X1=X1+1
2190 IF T8=0 THEN 2890
2200 IF T8=2 THEN 2400
2210 IF T8=3 THEN 2640
2220 B1=A(I1,1)+A(1,J1)
2230 W(X1)=B1
2240 IF X1<3 THEN 1510
2250 B2=W(X1)-W(X1-1)
2260 B3=W(X1)-W(X1-2)
2270 IF B2=0 THEN 2300
2280 IF B3=0 THEN 2300
2290 GO TO 1510
2300 H3=D1+D2
2310 X(2,1)=T8
2320 X(2,2)=H3
2330 X(2,3)=R3
2340 X(2,4)=I1
2350 X(2,5)=J1
2360 X(2,6)=C2
2370 X(2,7)=A1
2380 X(2,8)=D2
2390 GO TO 3230
2400 H1=D1+D2
2410 H3=0
2420 H1=H1-10
2430 H3=H3+10
2440 IF H1>0 THEN 2420
2450 IF H3>=L1*1000/4 THEN 4710
2460 B1=H3+A(I1,1)+A(1,J1)
2470 W(X1)=B1
2480 IF X1<3 THEN 1490
2490 B2=W(X1)-W(X1-2)
```

```
2500 B3=W(X1)-W(X1-1)
2510 IF B2=0 THEN 2540
2520 IF B3=0 THEN 2540
2530 GO TO 1490
2540 X(3,1)=T8
2550 X(3,2)=H3
2560 X(3,3)=R3
2570 X(3,4)=I1
2580 X(3,5)=J1
2590 X(3,6)=C2
2600 X(3,7)=A1
2610 X(3,8)=D2
2620 U=3
2630 GO TO 3490
2640 H1=D5+D2
2650 H3=0
2660 H1=H1-10
2670 H3=H3+10
2680 IF H1>0 THEN 2660
2690 M3=((1.6*W1*L1↑2)/8+(1.4*H3*W3*L1↑2)/8000)*1000000
2700 B8=1-(P1*F1)/(92*F2)
2710 IF B8<0.95 THEN 2730
2720 B8=0.95
2730 M2=F1*P1*10*(H3-D2)↑2*B8/1.15
2740 M9=0.15*F2*1000*(H3-D2)↑2
2750 IF M9<M2 THEN 2780
2760 M8=M2
2770 GO TO 2790
2780 M8=M9
2790 IF M8<M3 THEN 3180
2800 X(4,1)=T8
2810 X(4,2)=H3
2820 X(4,3)=R3
2830 X(4,4)=I1
2840 X(4,5)=J1
2850 X(4,6)=C2
2860 X(4,7)=A1
2870 X(4,8)=D2
2880 GO TO 3480
2890 H1=D1+D2
2900 H3=0
2910 H1=H1-10
2920 H3=H3+10
2930 IF H1>0 THEN 2910
2940 W4=H3*W3/1000
2950 W2=W4
2960 B1=H3+A(I1,1)+A(1,J1)
2970 W(X1)=B1
2980 IF X1=1 THEN 3050
2990 IF X1<3 THEN 1390
3000 B2=W(X1)-W(X1-2)
3010 B3=W(X1)-W(X1-1)
3020 IF B2=0 THEN 3070
3030 IF B3=0 THEN 3070
3040 GO TO 1390
3050 IF H3-T9>0 THEN 1390
3060 GO TO 1430
3070 X(1,1)=T8
3080 X(1,2)=H3
3090 X(1,3)=R3
3100 X(1,4)=I1
3110 X(1,5)=J1
3120 X(1,6)=C2
3130 X(1,7)=A1
3140 X(1,8)=D2
3150 GO TO 3210
```

```
3160  T8=3
3170  T=0
3180  T=T+1
3190  P1=0.25+(T-1)*0.2
3200  GO TO 3240
3210  P1=X(1,3)*X(1,7)/(10*(X(1,2)-X(1,8)))
3220  GO TO 3240
3230  P1=X(2,3)*X(2,7)/(10*(X(2,2)-X(2,8)))
3240  FOR U=2 TO 6
3250  IF M(U,1)=F1 THEN 3270
3260  NEXT U
3270  FOR V=2 TO 9
3280  IF M(1,2)>=P1 THEN 3320
3290  IF M(1,9)<=P1 THEN 3340
3300  IF M(1,V)>P1 THEN 3360
3310  NEXT V
3320  P5=M(U,2)
3330  GO TO 3410
3340  P5=M(U,9)
3350  GO TO 3410
3360  Q1=M(1,V)
3370  Q2=M(1,V-1)
3380  Q3=M(U,V)
3390  Q4=M(U,V-1)
3400  P5=(Q1-P1)*(Q4-Q3)/(Q1-Q2)+Q3
3410  IF L1>10 THEN 3440
3420  O1=20
3430  GO TO 3450
3440  O1=30-L1
3450  D5=L1*1000/(O1*P5)
3460  IF T8=3 THEN 1610
3470  IF D5>X(T8+1,2)-X(T8+1,8) THEN 3160
3480  U=T8+1
3490  P2=50*X(U,7)*X(U,3)/(1000*(X(U,2)-X(U,8)))
3500  FOR J=2 TO 5
3510  IF F2=V(1,J) THEN 3530
3520  NEXT J
3530  V8=V(7,J)
3540  FOR I=2 TO 6
3550  IF V(2,1)>=P2 THEN 3590
3560  IF P2>=V(6,1) THEN 3610
3570  IF V(I,1)>P2 THEN 3630
3580  NEXT I
3590  V9=V(2,J)
3600  GO TO 3680
3610  V9=V(6,J)
3620  GO TO 3680
3630  Q1=V(I,J)
3640  Q2=V(I-1,J)
3650  Q3=V(I,1)
3660  Q4=V(I-1,1)
3670  V9=(P2-Q4)*(Q1-Q2)/(Q3-Q4)+Q2
3680  FOR I=1 TO 5
3690  IF H3>=R(1,1) THEN 3730
3700  IF H3<=R(5,1) THEN 3750
3710  IF R(I,1)<H3 THEN 3770
3720  NEXT I
3730  R9=R(1,2)
3740  GO TO 3820
3750  R9=R(5,2)
3760  GO TO 3820
3770  Q1=R(I,2)
3780  Q2=R(I-1,2)
3790  Q3=R(I,1)
3800  Q4=R(I-1,1)
3810  R9=(Q4-H3)*(Q1-Q2)/(Q4-Q3)+Q2
```

```
3820 V4=L1*(1.6*W1+(1.4*X(U,2)*W3/1000))/(2*(X(U,2)-X(U,8)))
3830 X(U,9)=V4
3840 X(U,10)=V9*R9
3850 IF T8=2 THEN 3950
3860 IF V4>V8 THEN 3890
3870 IF V4>R9*V9 THEN 3950
3880 GO TO 3920
3890 T9=V4*X(U,2)/V8+X(U,8)
3900 D2=X(U,8)
3910 GO TO 1460
3920 IF T8=0 THEN 4510
3930 IF T8=1 THEN 4530
3940 IF T8=3 THEN 4570
3950 FOR E=1 TO 7
3960 B9=B(1,E)↑2*3.142/4
3970 B(2,E)=0.87*F1*B9/(X(U,9)-X(U,10))
3990 B(3,E)=SQR(B(2,E))
4000 B(4,E)=2*B(3,E)
4010 B(5,E)=B(4,E)/4
4020 NEXT E
4030 FOR F=3 TO 5
4040 FOR E=1 TO 7
4050 IF B(F,E)<100 THEN 4080
4060 IF B(F,E)>X(U,2)-X(U,8) THEN 4080
4070 GO TO 4100
4080 B(F,E)=0
4090 GO TO 4150
4100 B7=0
4110 B(F,E)=B(F,E)-10
4120 B7=B7+10
4130 IF B(F,E)>10 THEN 4110
4140 B(F,E)=B7
4150 NEXT E
4160 NEXT F
4170 FOR E=1 TO 7
4180 IF B(4,E)>0 THEN 4200
4190 B(5,E)=0
4200  IF B(5,E)>0 THEN 4220
4210 B(4,E)=0
4220 NEXT E
4230 S=0
4240 FOR E=1 TO 7
4250 S=S+B(3,E)+B(4,E)+B(5,E)
4260 NEXT E
4270 IF S>0 THEN 4330
4280 PRINT
4290 PRINT "PROVIDE THE FOLLOWING MINIMUM SHEAR REINFORCEMENT,"
4300 PRINT
4310 PRINT "8MM BARS AT"B8"MM X"4*B8"MM CRS,"
4320 GO TO 4460
4330 PRINT
4340 PRINT "CHOOSE VERTICAL SHEAR REINFORCEMENT FROM"
4350 PRINT "THE FOLLOWING ARRANGEMENTS:-"
4360 PRINT
4370 PRINT
4380 FOR E=1 TO 7
4390 IF B(3,E)=0 THEN 4410
4400 PRINT B(1,E)"MM BARS AT"B(3,E)"MM X"B(3,E)"MM CRS,"
4410 NEXT E
4420 FOR E=1 TO 7
4430 IF B(4,E)=0 THEN 4450
4440 PRINT B(1,E)"MM BARS AT"B(4,E)"MM X"B(5,E)"MM CRS,"
4450 NEXT E
4460 PRINT
4470 PRINT
4480 IF T8=1 THEN 4530
```

```
4490 IF T8=2 THEN 4550
4500 IF T8=3 THEN 4570
4510 PRINT "BALANCED DESIGN cRITERION"
4520 GO TO 4580
4530 PRINT "MINIMUM DEPTH CRITERION"
4540 GO TO 4580
4550 PRINT "SHEAR CRITERION"
4560 GO TO 4580
4570 PRINT "DEFLECTION CRITERION"
4580 PRINT "SLAB DEPTH="X(U,2)"MM"
4590 PRINT "NUMBER OF LAYERS OF REINFORCEMENT="X(U,3)
4600 PRINT "BAR DIAMETER="A(X(U,4),1)"MM"
4610 PRINT "BAR CENTRES="A(1,X(U,5))"MM"
4620 PRINT "COVER TO MAIN REINFORCEMENT="X(U,6)"MM"
4630 IF R3=1 THEN 4650
4640 PRINT "DISTANCE BETWEEN LAYERS=20MM"
4650 PRINT "CALCULATED STEEL AREA="X(U,3)*X(U,7)"MM↑2"
4660 PRINT "STEEL AREA PROVIDED="X(U,3)*A(X(U,4),X(U,5))"MM↑2"
4670 PRINT
4680 PRINT
4690 PRINT
4700 GO TO 4730
4710 PRINT "DESIGN DEPTH > SPAN/4 - USE DEEP BEAM THEORY."
4720 PRINT
4730 GO TO 870
4740 END
```

2.2.4.1 Block (A): Statements 100–770

In this block design information which is common to all problems is read into the computer. This comprises seven arrays of information called A( ), B( ), R( ), F( ), V( ), S( ) and M( ) respectively. There is no significance in the order in which these arrays appear in the program.

Array A( ) holds information concerning bar areas. Since this is a 10×23 element array the DIM statement at line 100 is necessary to ensure that sufficient storage space is set aside. It is simpler to calculate the 198 areas of steel included in this array each time the program is used than to type them out in the form of permanent data. The procedure for this takes the following form. The statements at lines 110 to 130 cause the bar centres to be entered into the first row of array A( ). These range from 75 mm in $A(1, 2)$ to 600 mm in $A(1, 23)$ at intervals of 25 mm. In a similar way nine bar diameters are entered into the first column of A( ) at lines 140 to 160. Thus elements $A(2, 1)$ to $A(10, 1)$ hold bar diameters ranging from 6 mm to 40 mm. For any element $A(I, J)$ the statements at lines 180 to 220 calculate the steel area in mm²/m for the bar diameter and centres represented by the row and column locations of the element. Thus the content of $A(4, 7)$ say, is 392.75, i.e. the area/metre width of 10 mm diameter bars at 200 mm centres.

The remaining arrays in this block represent the contents of various design tables from CP 110 : 1972. These will be referred to at a later stage when their roles in the design calculation are discussed.

2.2.4.2 Blocks (B) and (C): Statements 780–1350

Block (B) is prefaced by the PRINT statements at lines 780 to 840 which remind the designer of the available grades of concrete and steel. No matter how many individual problems are solved during one run of the program this information is only printed out once and is thereafter bypassed.

During the course of a solution intermediate results to which reference must be made at a later stage in a design iteration are stored in a 4×10 element array called $X(\ )$. At the beginning of the second and subsequent iterations this array must be cleared of the information which it already holds. This operation is carried out between lines 880 and 920.

A primary role of Block (B) is to present the computer with information which exactly defines the problem to be solved. These basic design parameters are:

the span in metres
the superimposed load in kN/m
the unit weight of concrete in kN/m^3
the minimum acceptable slab depth in mm
the characteristic steel strength in N/mm^2
and the characteristic concrete strength in N/mm^2

This information is called for and read into the computer at lines 950 to 990.

A minimum acceptable slab depth would normally be used to force a solution into never giving a depth less than one of minimum practical dimensions. By overestimating the minimum acceptable slab depth the designer could also use this facility to speed up convergence in cases where he is confident that deep slabs would result.

According to the program specification given in Section 2.2.1 spans are limited to 20 m. If a span greater than 20 m is entered by the designer in error then this fact is recognized at line 1010 and a new set of data must be submitted.

Block (C) comprises a single statement at line 1050. It constitutes the means whereby the program run is terminated. If in reply to a request for information at line 950 the designer types in a string of six zeros separated by commas the computer will recognize that since the span (L1) is zero the program run must be terminated. This is effected by making a jump to the last line in the program.

The exposure condition is set by typing in a number between 1 and 4 at line 1100. The minimum allowable concrete cover to reinforcement depends, amongst other things, on the concrete quality and exposure condition. Indeed, for one or more of the exposure conditions concrete exhibiting the lower characteristic concrete strengths may not be used. For this reason, even though the designer specifies a concrete strength at line 990, a check is made within the program to ascertain whether or not that concrete strength may be used in the specified exposure situation. If it cannot be used then the minimum concrete strength for that degree of exposure is printed out and if accepted by the designer at line 1250 the calculation proceeds. The operations outlined above takes place between lines 1070 and 1300. In the following paragraph they are considered in more detail.

F( ) is a 5×5 element array which holds in its first row a list of concrete strengths. The last four elements in a given column hold the recommended concrete covers for increasingly severe exposure conditions. Specifying row and column locations by X and Y respectively, if $F(X, Y)$ is zero then the concrete strength represented by $F(I, Y)$ may not be used in the exposure conditions represented by that value of X. The statements at lines 1120 to 1140 locate the chosen characteristic concrete strength and at line 1150 this leads to a value of the minimum cover, C2, for the actual exposure condition. If this is greater than zero then the calculation proceeds to line 1310. Otherwise at lines 1170 to 1190 a search is made along that row to locate the first cover which is greater than zero. Its position in the row defines the minimum concrete strength which may be used in conjunction with the current exposure condition. If this concrete strength is acceptable then the calculation proceeds, otherwise a revised set of data must be presented at line 990.

Although there are generally nine different bar diameters available it is useful to have a facility which allows a limit to be placed upon the number of diameters from which the computer may make a choice. Some bar diameters may be preferred because they are cheaper per unit weight than others. Some sizes may have to be excluded due to non-availability and in any case the designer will naturally want to limit the number of bar sizes used on any one job. Such limitations to choice may be imposed on the design by using the preferred diameter array P( ). This is a linear, nine element array. A reminder of the available bar diameters is printed out at line 1310 and the preferred diameters are input at line 1340. This data takes the form of a string of 1s and 0s, separated by commas. If $P(X) = 1$, then the bar diameter implied by the value of X is considered in the design calculation. Otherwise, if $P(X) = 0$, then the corresponding bar diameter is ignored.

2.2.4.3 Block (D): Statements 1360–1420

In this block the effective slab depth and steel area required by a 'balanced' design are determined. The switch (T8) is therefore set to zero at line 1360. It will be appreciated that because the self-weight of the slab comprises a significant proportion of the total loading the final design must be based upon an accurate weight assessment. At the beginning of the first iteration however no accurate information is available concerning the final self-weight. As a first approximation the initial weight is taken to be that of the slab of minimum acceptable depth. Hence the first slab weight assessment calculated at line 1380 is based upon T9. Using a bending moment based upon this approximate weight to begin with, and updated values later, the effective slab depth and steel area are calculated at lines 1400 and 1410 respectively. At the beginning of the first iteration the computer does not 'know' upon which criterion the final design will be based, however, checks which establish the actual criterion are made later in the calculation.

2.2.4.4 Block (E): Statements 1610–1870

In this block the bar diameter, number of rows and area per row are determined for a given area of steel.

At lines 1640 to 1670 the location of the largest preferred diameter in array P( ) is found and copied into P9. At line 1680 the required area A1 is compared with that area in array A( ) given by the largest preferred diameter at its closest centres, A(P9 + 1, 2). If A1 is less than this, then only one row of reinforcement is needed and R3 therefore remains equal to 1, the value it was initially assigned at line 1610. Otherwise R3 is increased by 1 and at line 1710 the area A1 now represents the area per row. This new value of A1 is compared with A(P9 + 1, 2) at line 1680, a process which continues until the area per row is less or equal to the largest preferred diameter at its closest centres. Thus the actual number of rows, R3, is now known. The row and column locations in A( ) of the chosen bar are copied into the variables I1 and J1 at lines 1840 and 1850 respectively. In this way, if it is found to be necessary at a later stage, all information pertaining to the chosen reinforcement may be recalled.

If a single layer of reinforcement is sufficient then the bar diameter and spacing chosen will be those which can provide the closest area which is just in excess of that required. This calculation is carried out between lines 1770 and 1870. Only a limited number of the array A( ) elements need to be examined. If a bar diameter appears as zero in the preferred diameter array P( ) then the relevant row in array A( ) is omitted from the search—see line 1790. Of the rows in A( ) which are examined, an upper limit to the bar centres (and hence to the distance along the row within which a solution is sought) is specified by the value of N7—see line 1780 (and line 2170 in Block (G)). The calculation consists of determining a quantity R1 which is the difference between the area held in element A(I, J) and the actual area required—see line 1800. If R1 is less than zero then the content of the next element is examined—see line 1810. At lines 1820 and 1830 R2 is a variable into which successively smaller values of R1 are copied. Coincidental with this values of I1 and J1 are recorded. When all the relevant elements of array A( ) have been examined the surviving values of I1 and J1 define the locations in A( ) holding information concerning the chosen reinforcement.

2.2.4.5 Block (F): Statements 1880–1900

The overall slab depth is the sum of the effective depth and the distance from the centre of gravity of the reinforcement to the tension face. In the program the latter dimension is called D2. It is a function of the bar diameter, the number of rows, the distance between each row and the cover to the reinforcement.

One assessment of the cover, C2, was made in Block (B). Another criterion which must be met is that C2 cannot be less than the diameter of the main reinforcement. The final choice of cover is made at lines 1880 and 1890.

D2 is calculated at line 1900 on the assumption that there is a distance of 20 mm between layers of reinforcement.

2.2.4.6 *Block (G) : Statements 1910–2210*

Under normal circumstances, if bars are placed at no greater centres than those advised in CP 110 the distribution and width of cracks will be limited to an acceptable degree. The maximum spacing of bars is controlled by the

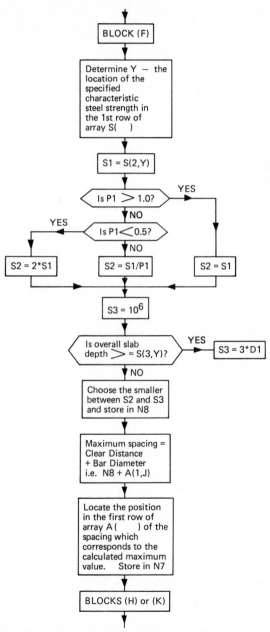

Figure 2.2 Flow diagram for block (G)—program RC4

characteristic strength of the steel, the steel percentage and the overall slab depth.

A 3×5 element array called S( ) contains the information on which to decide the maximum bar spacing for a particular problem. The first row holds five characteristic steel strengths and the second, the allowable clear horizontal distances between bars (for zero redistribution of bending moment at the section considered) appropriate to the contents of the first row. These two rows comprise the relevant part of Table 24 CP 110 for simply supported slabs. The third row of array S( ) holds the maximum overall slab thicknesses appropriate to the reinforcement strengths, for which a maximum clear distance between bars of three times the effective depth is allowed.

The operations carried out in Block (G) are shown in the flow diagram in Fig. 2.2. The first stage, carried out at lines 1940 to 1960, locates the position (Y) of the specified steel strength (F1) in the first row of array S( ). The allowable bar spacing is therefore S(2, Y) and this quantity is transferred to the variable S1 at line 1970. Since the acceptable spacing is a function of the steel percentage (P1) S1 is factored to meet this requirement between lines 1980 and 2030. The result is transferred to the variable S2 at line 2040.

The slab depth bar spacing criterion is now considered. If the slab depth is not greater than that held in S(3, Y), S3 becomes equal to three times the effective depth. Otherwise S3 remains at the artificially inflated value it was given at line 2050. The smaller of the two values S2 or S3 is then chosen and the result is placed in the variable N8—see lines 2090 to 2120. At line 2130 the permissible maximum spacing becomes equal to (N8 + the chosen bar diameter).

A search is made through the first row of array A( ) at lines 2140 to 2160 to locate the value of J which corresponds with the first available spacing in excess of N8. A variable called N7 is then set to (J–1) at line 2170. In the next iteration this value of N7 is used at line 1780 to limit the scope within which a steel area is chosen.

The actual position of Block (G) in the program is inefficient, at least in respect of the first design iteration. Its location was determined by the fact that one of the bar spacing criteria depended upon a knowledge of the overall slab depth. The bar spacing calculation was therefore postponed until the depth was known—even though to some degree D2 (and therefore the overall slab depth) is in turn affected by the bar spacing. This 'chicken and egg' dilemma often arises in design. In manual design it can often be resolved by intelligent anticipation. In computer aided design such inconsistencies can be overcome by carrying out a sufficient number of iterations.

2.2.4.7 Blocks (H) and (I): Statements 2890–3150

This is a program area which is entered by all problems during the first iteration and subsequently only by those for which the bending solution is given by 'balanced' design.

At this stage a theoretical overall depth is known together with details of reinforcement. A solution is being achieved iteratively by using the previous solution to give another one which is even closer to what is considered acceptable. The results of alternate and consecutive solutions are compared in this block to decide whether or not the design criterion has been met.

For practical purposes it is necessary to round-up the theoretical value of the overall depth to the nearest 10 mm. This happens at lines 2890 to 2930. The slab weight (W2) is calculated at lines 2940 and 2950 to be used in the next iteration should one be necessary. A slab solution is uniquely defined by summing the overall depth, the bar diameter and the bar spacing. This quantity, called B1 in the program, is calculated at line 2960. The value of B1 is transferred to a linear array $W(\ )$ at line 2970, where it is held in the element which is defined by the number of the current iteration $(X1)$. At the end of the first iteration a comparison is made at line 3050 between the 'balanced' design depth and the minimum practical depth. If at this stage H3 is less than T9 there is little point in determining a more accurate value of H3 because it will always be less than T9. The route is therefore switched at line 3060 to Block (J). If H3 is greater than T9 then a minimum of three full iterations are carried out before solutions are able to be compared at lines 3000 to 3030. These checks are made on the difference between the contents of alternate and consecutive elements in array $W(\ )$. If the checks show that either of these pairs of solutions are identical then the calculation proceeds to Block (L) after basic information concerning the solution is stored in the first row of array $X(\ )$ at lines 3070 to 3140.

2.2.4.8 Block (J): Statements 1430–1600

This block is entered at one of two points depending upon whether a minimum depth calculation or a shear depth calculation is being made. For the former, entry is at line 1430 where T8 is made equal to 1. For the latter the entry point is line 1460 where T8 is made equal to 2. Before entering the block it had already been shown that the effective depth required was greater than that for 'balanced' design. The current calculations are therefore those for an under-reinforced' design. The steel area A1 is initially calculated at line 1550 and the lever arm factor B8 at line 1560. If the lever arm factor is found to be greater than 0.95 then the steel area is reassessed at line 1580. If the calculated steel area is less than 0.25% then it is increased to this minimum value at line 1600.

2.2.4.9 Block (K): Statements 2220–2390

The function of this block is to determine when an acceptable design has been attained in the case of a slab having the minimum specified depth. Entry into the block is made at line 2220 where the variable B1 is set to the sum of the current trial bar diameter and its spacing, a quantity which uniquely defines the current solution. Successive iterations give values of B1 which are entered into the elements of array W( ) at line 2230. A minimum of three iterations are executed before comparisons are made between alternate and consecutive solutions at lines 2250 to 2280. When either comparison shows that a pair of designs are identical, details of that slab are entered into the second row of array X( ). Otherwise the statement at line 2290 directs that another iteration will be carried out.

2.2.4.10 Blocks (L) and (R): Statements 3160–3470

CP 110 makes recommendations which limit the span/effective depth ratio to a maximum of 20 for spans of up to 10 m and thereafter a ratio which reduces linearly to 10 as spans increase to 20 m. Further recommendations are made concerning the way in which deflections are affected by the amount of tension reinforcement at the centre of the span and its service stress. Based upon these factors, Table 10 CP 110 gives coefficients ranging from 0.68 to 2.0 by which the recommended span/effective depth ratios are to be factored. An edited version of this table is held in array M( ).

For the problems solved by this program there is no redistribution of bending moment at mid-span; in addition to this the reinforcement provided to resist bending is effectively the same as that calculated. It is therefore assumed that the service stress is equal to $0.58f_y$. Hence only those rows in Table 10 CP 110 corresponding to a service stress of $0.58f_y$ are stored together with a first row which holds steel percentages ranging from 0.25 to 3.0. This array, called M( ), therefore has 6×9 elements.

Lines 3240 to 3450 are common to both the Blocks (L) and (R). In the case of Block (L) they form part of a deflection check calculation; when functioning as Block (R) they are used to create a new section which will satisfy the deflection criterion.

There are two entry points to Block (L); either at line 3210 if $T8 = 0$ or at line 3230 if $T8 = 1$. At each of these lines the appropriate steel percentage P1 is calculated for the largest current section on the basis of the section information stored in array X( ).

The first task is to locate the modification coefficient in array M( ) which relates to the steel percentage and service stress (or characteristic steel strength since these may be interchanged under the design conditions assumed). The row U, which holds the specified characteristic steel strength, is located by the statements at lines 3240 to 3260. At lines 3270 to 3310 the first row of M( ) is searched to find a steel percentage which is just in excess of the section steel percentage. At this stage known values of U and V locate the element $M(U, V)$ which holds a modification coefficient just smaller than the one required. By the same token $M(U, V - 1)$ will be just too large. The actual value (called P5) is found by interpolating between these two at lines 3320 to 3400.

The allowable span/depth ratio O1 is determined at lines 3410 to 3440. This calculation takes account of the alternative criteria to be applied when the span is greater or less than 10 m. Now that O1 and P5 have been set to known values the minimum effective depth necessary to meet the deflection criterion (D5) is found at line 3450. A comparison is made at line 3470 between the depth required for deflection and that provided by the current section. If the depth provided is greater than that required then the calculation proceeds to Block (M). Otherwise Block (R) is entered at line 3160 and T8 is made equal to 3.

What was previously a check calculation now becomes more of a formal design calculation—one to derive a section which will satisfy both the bending and deflection criteria. The routine begins by specifying a steel percentage of 0.25 at line 3190 and thereafter determining the effective slab depth D5 which meets deflection requirements when associated with this steel percentage. The statement at line 3460 causes the calculation to jump back to the beginning of Block (E)—followed naturally by Blocks (F), (G) and (S). At Block (S) the slab is checked against bending requirements and if found to be too small the calculation re-enters Block (R) at line 3180. At line 3190, P1 is arbitrarily advanced by 0.2%—this leads to a smaller modification factor and hence a greater depth of slab—and the whole routine is repeated. The cycle continues until a steel percentage is found which simultaneously satisfies the deflection criterion at Block (R) and the bending criterion at Block (S).

2.2.4.11 Block (S): Statements 2640–2880

The function of Block (S) was referred to in the previous section. This block is entered at line 2640 where the first operation is to round-up the theoretical value of the overall depth H1 to the nearest 10 mm and to put this result into the variable H3. The applied bending moment (M3) is calculated at line 2690. The moment of resistance of the steel (M2) and the concrete (M9) are determined at lines 2730 and 2740 respectively. The smaller of these is compared with M3 at line 2790. If the internal moment of resistance of the section is less than the applied bending moment then the calculation must remain in the deflection criterion iteration. This it does by returning to Block (R) at line 3180. Otherwise the section is acceptable on both bending and deflection counts and relevant section information is recorded in the fourth row of array X( ). The calculation then moves on to the shear checks at Block (M).

2.2.4.12 Block (M): Statements 3480–3850

The purpose of this block is to determine, for the deepest slab so far produced, the actual shear stress (V4) and the allowable shear stress (V9*R9).

The allowable shear stress is a function of the percentage of steel at the support, the characteristic concrete strength and the overall slab depth. The variation of allowable shear stress with the first two of these parameters is given in Table 5 CP 110. In the program, this information is stored in the first six rows of a 7×5 array called V( )—see lines 520 to 570. The effect of overall slab depth upon the allowable shear stress is introduced by applying the variable factor given in Table 14 CP 110. This information is stored in a 5×2 element array called R( )—see lines 320 to 360.

In addition to an allowable shear stress, which if exceeded requires the provision of shear reinforcement, there is a level of shear stress which must never be exceeded. In practice it is often found that designing up to this limit results in difficulties with the placing of shear reinforcement. For this reason the recommended absolute maximum shear stress levels have been reduced by 33%. These maximum stress values, related to the appropriate characteristic concrete strengths in the first row of array V( ), are stored in its seventh row— see line 580.

Block (M) is entered at line 3480. The value of T8 on entering the block determines the design criterion which up to this stage has governed the design— 'balanced', minimum depth or design for deflection. Moreover, since T8 is related to the row number in array X( ) in which information concerning this design is stored, then P2 (the steel percentage at the support) is readily calculated at line 3490.

At lines 3500 to 3520 a search through the first row of array V( ) locates the element $V(1, J)$ which holds the specified characteristic concrete strength. Therefore the absolute maximum allowable shear stress for this grade of concrete (V8) is known to be held in $V(7, J)$—see line 3530. A search is made through the first column of V( ) to find the first steel percentage which is greater than P2—see lines 3540 to 3580. This is located in row I. Element $V(I, J)$ therefore holds a shear stress which is just in excess of the allowable value. V9, the allowable shear stress, is found by interpolating between the contents of $V(I, J)$ and $V(I, J - 1)$ at lines 3590 to 3670.

A similar type of operation is carried out in array R( ) to determine a value for R9, the variable factor which introduces the effect of slab depth—see lines 3680 to 3720.

At lines 3830 and 3840 the values of V4 and V9∗R9 are stored in the appropriate line of array X( ) for future reference.

2.2.4.13 Blocks (N), (O) and (T): Statements 3860–3940

The actual shear stress (V4) and the absolute maximum allowable shear stress value (V8) are compared at line 3860. If the maximum allowable shear stress level has been violated then an approximate slab depth calculated at line 3890 reduces V4 to the level of V8. Whilst shear reinforcement is still required its design is deferred until the slab depth for this condition has been finalized— a procedure which is initiated at line 3910 where the calculation is directed to Block (J) (see Section 2.2.4.8).

At line 3870, V4 is compared with the allowable shear stress (R9∗V9). If V4 is greater than that allowed then shear reinforcement is necessary and the calculation is directed to Block (P). Otherwise the statements at lines 3920 to 3940 determine the form in which the results will be output.

2.2.4.14 Block (U): Statements 2400–2630

This block of statements is concerned with recognizing when a slab, which has been primarily proportioned to satisfy the criterion of absolute maximum shear stress, is satisfactory in bending. The block is entered at line 2400. The theoretical overall depth H1 is rounded up at lines 2400 to 2440 to give an overall design depth H3. The quantity represented by B1 at line 2450 (i.e. the sum of the overall slab depth, the bar diameter and bar spacing) uniquely defines the current solution in this case. In other respects the remainder of this block takes an almost identical form to that of Block (K)—see Section 2.2.4.9 for details.

2.2.4.15 Block (P): Statements 3950–4270

If this block is entered it is understood that shear reinforcement is required. The outcome of this calculation is to produce the necessary spacing of vertical reinforcement for a range of bar sizes. These reinforcements would probably take the form of binders, in which case once the spacing was determined the binder width could be calculated. Some results are automatically suppressed by the program, in particular those spacings less than 100 mm or greater than the effective depth of the slab. Even so, the shear reinforcement output is only meant to indicate theoretical possiblities. It remains with the designer to choose a practical arrangement from amongst those suggested.

Seven bar diameters, ranging from 6 mm to 25 mm, were read into the first row of a 5×7 element array called B( ) at lines 230 to 260. The primary function of this array is to hold information concerning the spacing of vertical shear reinforcement.

The tensile force which is generated over an area of concrete (A) by a shear stress which exceeds that allowable by an amount (V) is $(A*V)$. The ultimate tensile force developed by a bar of area (B) is $(0.87*f_y*B)$. Equating these two forces gives $A = 0.87*f_y*B/V$. (A) can be thought of as being that area which is served by a single bar. In the program it is expressed either as a square of side $\sqrt{A}$ or a rectangle of sides $2\sqrt{A}$ and $\sqrt{A/2}$; these are also ways of expressing the required spacing of a bar of area (B). The programmed equivalents of (B, V and A) are:

(B) is the equivalent of the content of variable B9 which is calculated at line 3960

(V) is equivalent to $X(U, 9) - X(U, 10)$, i.e. the difference between V4 and $V9*R9$ (see lines 3830 and 3840)

(A) is the content of $B(2, E)$ which is calculated at line 3970.

For each of the seven diameters held in B( ) their spacings as shear reinforcement are calculated at lines 3950 to 4020. For a bar of diameter $B(1, E)$ its spacing according to a 'square' criterion is entered into $B(3, E)$; 'rectangular' criterion spacings are located in $B(4, E)$ and $B(5, E)$ respectively.

At lines 4030 to 4160 the two following operations are carried out on the contents of the elements in the last three rows of array B( ): any element whose content is either less than 100 mm or greater than the effective depth is made equal to zero; the remaining elements are rounded down to the nearest 10 mm.

At lines 4170 to 4220 corresponding pairs of elements in rows 4 and 5 of B( ) are examined; if the content of an element in either row is zero then the corresponding element in the other row is set to zero.

The final task, at lines 4230 to 4260, is to sum the contents of all the elements in the last three rows of B( ). The significance of this is that whilst all the slabs considered in this block require shear reinforcement the calculated reinforcement could be so light that it is defined theoretically by the smallest bar diameter at centres which are greater than the effective depth of the slab. The sum (S) is checked at line 4270; if it does equal zero then minimum shear reinforcement is suggested at line 4290.

2.2.4.16 Block (Q): Statements 4280–4740

This block is primarily concerned with the output of results. If shear reinforcement is required then the results are prefaced by a statement of possible vertical bar arrangements. The main body of the results follow with details of the slab and its bending reinforcement headed by a statement of the criterion on which the design is based.

The statement at line 4710 prints out a 'design fail' message if the design depth is greater than Span/4, a depth beyond which it is assumed that normal beam behaviour cannot be expected.

Regardless of the outcome of a design the GOTO statement at line 4730 returns the program user to NEXT PROBLEM at line 870.

2.2.4.17 Example of Program RC4 Output

A typical solution given by this program appears below. The single number, or string of numbers, following each question mark constitute the designer's response to a request by the computer for information. In this example the designer specified Grade 20 concrete and Exposure Condition 3. But since these two parameters are incompatible the computer replied that the minimum acceptable concrete grade for this exposure condition was 25. The designer accepted this advice and went on to specify that the reinforcement should be chosen from the 20 mm, 25 mm and 32 mm diameter group of bars. This completed the input of design parameters. The computer then executed the main body of the design calculation and, for this example, gave a solution which was based upon the deflection criterion.

```
RC4

CONCRETE GRADE (FCU) TO BE CHOSEN FROM THE FOLLOWING
20,25,30,40 AND 50 N/MM↑2
```

```
STEEL GRADE (FY) TO BE CHOSEN FROM THE FOLLOWING
250,410,425,460 AND 500 N/MM†2

SPAN(M),SUPERLOAD(KN/M†2),CONCRETE WEIGHT(KN/M†3),

MINIMUM SLAB DEPTH(MM),FY(N/MM†2),FCU(N/MM†2).

 ? 8.2,7,24,100,250,20


TYPE 1,2,3 OR 4 FOR MILD,MODERATE,SEVERE OR VERY SEVERE

EXPOSURE CONDITIONS ? 3

MINIMUM FCU FOR THESE EXPOSURE CONDITIONS= 25 N/MM†2

IF THIS VALUE OF FCU IS ACCEPTABLE TYPE 1 ELSE 0 ? 1

AVAILABLE BAR DIAMETERS ARE 6,8,10,12,16,20,25,32,40MM

PREFERRED DIAMETERS ? 0,0,0,0,0,1,1,1,0

DEFLECTION CRITERION
SLAB DEPTH= 370 MM
NUMBER OF LAYERS OF REINFORCEMENT= 1
BAR DIAMETER= 25 MM
BAR CENTRES= 125 MM
COVER TO MAIN REINFORCEMENT= 50 MM
CALCULATED STEEL AREA= 3824.63 MM†2
STEEL AREA PROVIDED= 3927.5 MM†2


NEXT PROBLEM

SPAN(M),SUPERLOAD(KN/M†2),CONCRETE WEIGHT(KN/M†3),

MINIMUM SLAB DEPTH(MM),FY(N/MM†2),FCU(N/MM†2).

 ? 0,0,0,0,0,0


RUNNING TIME:   7.3 SECS   I/O TIME :   5.3 SECS
```

2.3 The Decision Design of Reinforced Concrete Slabs

2.3.1 Program Specification—RC9

The purpose of this program is to determine, on the basis of input designer decisions, the slab depth and reinforcement required for simply supported, singly reinforced slabs of up to 20 m span, for conditions of mild exposure. The slabs are designed for the ultimate limit state of bending and the serviceability limit states of deflection and cracking as defined in CP 110:1972. The designer is informed if shear reinforcement is required but no facility for calculating this is provided within the program.

2.3.2 The RC9 Flow Diagram

This flow diagram is shown in Fig. 2.3; individual blocks of program are again identified by letters. The function of blocks having identifying letters between (A) and (Q) is the same in some cases, and almost identical in others, as similarly identified blocks of program in RC4.

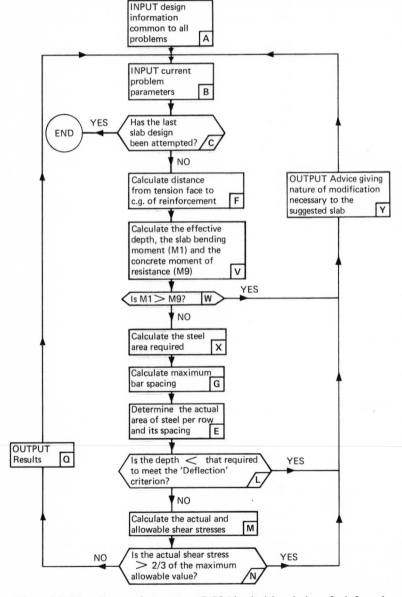

Figure 2.3 Flow diagram for program RC9 (the decision design of reinforced concrete slabs)

Program RC9 was written to perform check calculations on a slab of known overall depth, and given the bar diameter, to design the bending reinforcement. At three stages in the calculation—those of bending, deflection and shear— whenever a check indicates a failure of the proposed slab to meet design requirements the designer is so informed and must increase the slab depth or bar diameter accordingly. A solution is therefore achieved iteratively, but the process is not an automatic one. In the search for an acceptable solution the designer is at liberty either to increase or decrease the slab depth and bar diameter; at his request results may be output for slabs of excessive strength but never for those of insufficient strength to meet CP 110 requirements.

For an input slab depth which meets all the programmed design criteria the route to a solution is a direct one which follows the path:

$$A \rightarrow B \rightarrow C \rightarrow F \rightarrow V \rightarrow W \rightarrow X \rightarrow G \rightarrow E \rightarrow L \rightarrow M \rightarrow N \rightarrow Q$$

A failure at either of the Blocks (W), (L) or (N) returns the calculation to Block (B) where updated information is requested. This simple route results from the fact that the program is dealing with a slab of fixed depth and known bar diameter and therefore at each stage the force actions and effective depth are accurately known.

2.3.3 Writing Program RC9

An edited version of RC4 provided the framework on which RC9 was built. The basic editing was carried out on-line. A temporary file copy of RC4 was renamed RC9 and, using the EDIT DELETE facility, irrelevant areas of the program were cancelled. The remainder, which was then listed, comprised a substantial part of the new program. The input and output statements were modified to suit the new approach and further linking statements and jumps were provided where necessary. The main reorganization of what remained consisted of moving Block (G) (concerned with the assessment of maximum bar spacing) to its proper position in the order of calculation. This was possible because, unlike Program RC4, in this program the final slab depth is known from the outset.

The variable identifiers and arrays used in Program RC9 are listed in Section 2.2.4 along with those of RC4.

2.3.4 Description of Program RC9

The program is listed below and should be read in conjunction with the flow diagram shown in Fig. 2.3.

```
100 DIM A(10,23)
110 FOR J=2 TO 23
120 A(1,J)=75+(J-2)*25
130 NEXI J
```

```
140 FOR I=2 TO 10
150 READ A(I,1)
160 NEXT I
170 DATA 6,8,10,12,16,20,25,32,40
180 FOR I=2 TO 10
190 FOR J=2 TO 23
200 A(I,J)=(250*3.142*A(I,1)+2)/A(1,J)
210 NEXT J
220 NEXT I
230 FOR X=1 TO 5
240 FOR Y=1 TO 2
250 READ R(X,Y)
260 NEXT Y
270 NEXT X
280 DATA 250,1.0
290 DATA 225,1.05
300 DATA 200,1.1
310 DATA 175,1.15
320 DATA 150,1.2
330 FOR X=1 TO 2
340 FOR Y=1 TO 5
350 READ F(X,Y)
360 NEXT Y
370 NEXT X
380 DATA 20,25,30,40,50
390 DATA 25,20,15,15,15
400 FOR X=1 TO 7
410 FOR Y=1 TO 5
420 READ V(X,Y)
430 NEXT Y
440 NEXT X
450 DATA 0,20,25,30,40
460 DATA 0.25,0.35,0.35,0.35,0.35
470 DATA 0.50,0.45,0.5,0.55,0.55
480 DATA 1.0,0.6,0.65,0.7,0.75
490 DATA 2.0,0.8,0.85,0.9,0.95
500 DATA 3.0,0.85,0.9,0.95,1.0
510 DATA 0,2.24,2.51,2.75,3.18
520 FOR X=1 TO 3
530 FOR Y=1 TO 5
540 READ S(X,Y)
550 NEXT Y
560 NEXT X
570 DATA 250,410,425,460,500,
580 DATA 300,185,180,165,150
590 DATA 250,250,250,200,200
600 FOR U=1 TO 6
610 FOR V=1 TO 9
620 READ M(U,V)
630 NEXT V
640 NEXT U
650 DATA 0,0.25,0.5,0.75,1.0,1.5,2.0,2.5,3.0
660 DATA 250,2.0,1.98,1.62,1.44,1.24,1.13,1.06,1.01
670 DATA 410,1.6,1.23,1.09,1.0,0.9,0.84,0.8,0.77
680 DATA 425,1.55,1.2,1.06,0.98,0.88,0.83,0.79,0.76
690 DATA 460,1.41,1.11,0.99,0.92,0.84,0.78,0.75,0.72
700 DATA 500,1.27,1.03,0.92,0.86,0.79,0.74,0.71,0.68
710 PRINT "CONCRETE GRADE (FCU) TO BE CHOSEN FROM THE FOLLOWING"
720 PRINT "20,25,30,40 AND 50 N/MM+2"
730 PRINT
740 PRINT "STEEL GRADE (FY) TO BE CHOSEN FROM THE FOLLOWING"
750 PRINT "250,410,425,460 AND 500 N/MM+2"
760 PRINT
770 GO TO 800
780 PRINT "NEXT PROBLEM"
790 PRINT
```

```
 800 PRINT "SPAN(M),SUPERLOAD(KN/M↑2),CONCRETE WEIGHT(KN/M↑3),"
 810 PRINT "FY(N/MM↑2),FCU(N/MM↑2)"
 820 INPUT L1,W1,W3,F1,F2
 830 IF L1<=20 THEN 860
 840 PRINT "SPAN GREATER THAN 20M."
 850 GO TO 800
 860 IF L1=0 THEN 2450
 870 PRINT "SLAB DEPTH(MM),BAR DIAMETER(MM),NO. OF ROWS"
 880 INPUT T9,B9,R3
 890 IF T9>=L1*1000/4 THEN 2420
 900 FOR Y=1 TO 5
 910 IF F(1,Y)=F2 THEN 930
 920 NEXT Y
 930 C2=F(2,Y)
 940 IF F(2,Y)>B9 THEN 960
 950 C2=B9
 960 D2=(20*(R3-1)/2)+(R3*B9/2)+C2
 970 D1=19-D2
 980 M1=(1.6*W1*L1↑2)/8+(1.4*T9*W3*L1↑2)/8000
 990 M9=0.15*F2*D1↑2/1000
1000 IF M1>M9 THEN 2060
1010 Z1=F1*D1/1.15
1020 Z2=3780.718*F1*F1*M1/F2
1030 Z3=F1*F1/(529*F2)
1040 A1=(Z1-SQR(Z1↑2-Z2))/Z3
1050 B8=1-(A1*F1)/(920*F2*D1)
1060 IF B8<0.95 THEN 1080
1070 A1=(M1*10↑6)/(F1*0.95*D1)
1080 IF A1>=2.5*D1 THEN 1100
1090 A1=2.5*D1
1100 A1=A1/R3
1110 P1=A1*R3/(10*D1)
1120 FOR Y=1 TO 5
1130 IF S(1,Y)=F1 THEN 1150
1140 NEXT Y
1150 S1=S(2,Y)
1160 IF P1>1.0 THEN 1220
1170 IF P1<0.5 THEN 1200
1180 S2=S1/P1
1190 GO TO 1230
1200 S2=S1*2
1210 GO TO 1230
1220 S2=S1
1230 S3=1000000
1240 IF T9<=S(3,Y) THEN 1260
1250 GO TO 1300
1260 S3=3*D1
1270 IF S2>S3 THEN 1300
1280 N8=S2
1290 GO TO 1320
1300 N8=S3
1310 N8=N8+B9
1320 FOR J=2 TO 23
1330 IF (N8-A(1,J))<0 THEN 1350
1340 NEXT J
1350 N7=J-1
1360 FOR I=2 TO 10
1370 IF A(I,1)=B9 THEN 1390
1380 NEXT I
1390 I1=I
1400 IF A1>A(I1,2) THEN 2140
1410 FOR J=2 TO N7
1420 IF A(I1,J)<A1 THEN 1440
1430 J1=J
1440 NEXT J
1450 FOR U=2 TO 6
```

```
1460 IF M(U,1)=F1 THEN 1480
1470 NEXT U
1480 FOR V=2 TO 9
1490 IF M(1,2)>=P1 THEN 1530
1500 IF M(1,9)<=P1 THEN 1550
1510 IF M(1,V)>P1 THEN 1570
1520 NEXT V
1530 P5=M(U,2)
1540 GO TO 1620
1550 P5=M(U,9)
1560 GO TO 1620
1570 Q1=M(1,V)
1580 Q2=M(1,V-1)
1590 Q3=M(U,V)
1600 Q4=M(U,V-1)
1610 P5=(Q1-P1)*(Q4-Q3)/(Q1-Q2)+Q3
1620 IF L1>10 THEN 1650
1630 O1=20
1640 GO TO 1660
1650 O1=30-L1
1660 D5=L1*1000/(O1*P5)
1670 IF D5>D1 THEN 2060
1680 P2=P1/2
1690 FOR J=2 TO 5
1700 IF F2=V(1,J) THEN 1720
1710 NEXT J
1720 V8=V(7,J)
1730 FOR I=2 TO 6
1740 IF V(2,1)>=P2 THEN 1780
1750 IF P2>=V(6,1) THEN 1800
1760 IF V(I,1)>P2 THEN 1820
1770 NEXT I
1780 V9=V(2,J)
1790 GO TO 1870
1800 V9=V(6,J)
1810 GO TO 1870
1820 Q1=V(I,J)
1830 Q2=V(I-1,J)
1840 Q3=V(I,1)
1850 Q4=V(I-1,1)
1860 V9=(P2-Q4)*(Q1-Q2)/(Q3-Q4)+Q2
1870 FOR I=1 TO 5
1880 IF T9>=R(1,1) THEN 1920
1890 IF T9<=R(5,1) THEN 1940
1900 IF R(I,1)<T9 THEN 1960
1910 NEXT I
1920 R9=R(1,2)
1930 GO TO 2010
1940 R9=R(5,2)
1950 GO TO 2010
1960 Q1=R(I,2)
1970 Q2=R(I-1,2)
1980 Q3=R(I,1)
1990 Q4=R(I-1,1)
2000 R9=(Q4-H3)*(Q1-Q2)/(Q4-Q3)+Q2
2010 V4=L1*(1.6*W1+(1.4*T9*W3/1000))/(2*D1)
2020 IF V4<=R9*V9 THEN 2170
2030 IF V4<=V8 THEN 2090
2040 GOTO 2060
2050 PRINT
2060 PRINT "INCREASE SLAB DEPTH"
2070 PRINT
2080 GO TO 870
2090 PRINT
2100 PRINT "SOLUTION POSSIBLE BUT SHEAR REINFORCEMENT REQUIRED."
2110 PRINT
```

```
2120 GO TO 2190
2130 PRINT
2140 PRINT "INCREASE BAR DIAMETER OR NO. OF ROWS"
2150 PRINT
2160 GO TO 870
2170 PRINT "SOLUTION POSSIBLE"
2180 PRINT
2190 PRINT "IF RESULT REQUIRED TYPE 1 ELSE 0"
2200 INPUT Q8
2210 PRINT
2220 IF Q8=1 THEN 2240
2230 GO TO 870
2240 PRINT "SLAB DEPTH="T9"MM"
2250 PRINT "NUMBER OF LAYERS OF REINFORCEMENT="R3
2260 PRINT "BAR DIAMETER="A(I1,1)"MM"
2270 PRINT "BAR CENTRES="A(1,J1)"MM"
2280 PRINT "COVER TO MAIN REINFORCEMENT="C2"MM"
2290 IF R3=1 THEN 2310
2300 PRINT "DISTANCE BETWEEN LAYERS=20MM"
2310 PRINT "CALCULATED STEEL AREA="R3*A1"MM+2"
2320 PRINT "STEEL AREA PROVIDED="R3*A(I1,J1)"MM+2"
2330 PRINT
2340 PRINT
2350 PRINT
2360 PRINT "IF SOLUTION IS ACCEPTABLE TYPE 1 ELSE 0";
2370 INPUT Q7
2380 PRINT
2390 PRINT
2400 IF Q7=0 THEN 870
2410 GO TO 2440
2420 PRINT "DESIGN DEPTH > SPAN/4 - USE DEEP BEAM THEORY."
2430 PRINT
2440 GO TO 780
2450 END
```

2.3.4.1 Block (A): Statements 100–700

Apart from the exclusion of array B( ) and some modifications to the contents of array F( ) this block is identical with that discussed in Section 2.2.4.1.

2.3.4.2 Blocks (B) and (C): Statements 710–890

The first entry point to Block (B) is at line 710. The block is prefaced by PRINT statements at lines 710 to 760 which remind the designer of the available concrete and steel grades and bar diameters. During a program run the block is subsequently entered at line 780 each time a new problem is to be solved. The parameters which define the current problem are requested and input at lines 800 to 820. In comparison with Program RC4 the only item omitted here is that of the minimum acceptable slab depth; the designer will obviously have this dimension in mind when he inputs a slab depth at line 880.

The statement at line 860 comprises Block (C) whose function is to provide an exit from the program.

At lines 870 and 880 the designer is required to make a guess at the solution by assessing the probable overall slab depth, the bar diameter to be used and the number of rows of reinforcement.

2.3.4.3 Block (F): Statements 900–960

Because in this instance solutions are limited to those for mild exposure conditions the determination of concrete cover is not so involved as the one in Program RC4. The cover calculation takes place at lines 900 to 950. It simply comprises a search through the first row of array F( ) to locate the specified characteristic concrete strength, and hence the appropriate minimum concrete cover in the second row; a choice of the largest dimension given either by this or the diameter of the reinforcement then determines the value of C2. D2, the distance between the tension face and the centre of gravity of the reinforcement, follows at line 960.

2.3.4.4 Block (V): Statements 970–990

Following the calculation of D1 at line 970 both the effective depth and the overall depth of the proposed slab are known precisely. This allows accurate assessments of the maximum applied bending moment (M1) and the maximum concrete moment of resistance of the slab (M9) to be made at lines 980 and 990 respectively.

2.3.4.5 Block (W): Statement 1000

This comprises the first section check. If the applied bending moment is greater than the moment of resistance of the concrete (on the basis of 'balanced' design), then the section depth must be increased; hence the jump to line 2060 where this action is advised.

2.3.4.6 Block (X): Statements 1010–1090

If the check at Block (W) shows that the concrete has a sufficiently high moment of resistance, then the next stage in the calculation is to determine the area of steel required. This is an 'under-reinforced' section calculation which is carried out at lines 1010 to 1040. The lever-arm factor (B8) is calculated at line 1050 and its magnitude is checked at line 1060. If it proves to be greater than 0.95, then the steel area is reassessed at line 1070 using a lever-arm factor equal to 0.95. A final check on the magnitude of the steel area at lines 1080 and 1090 ensures that a minimum of 0.25% is adopted.

2.3.4.7 Block (G): Statements 1100–1350

The maximum allowable bar spacing is determined in an identical way to that described in Section 2.2.4.6.

2.3.4.8 Block (E): Statements 1360–1440

The purpose of this block is to determine the spacing of the designer specified bar diameter (B9) which gives an area per metre width equivalent to A1.

This routine is much simpler than the one described in Section 2.2.4.4 because the bar diameter to be used is known from the outset. Its location (I1) in the first column of array A( ) is found at lines 1360 to 1390. I1 now defines the row of elements within which an equivalent area will be sought. The largest area in the row, A(I1,2), is compared with A1 at line 1400. If it is smaller than A1 then a jump to line 2140 advises the designer either to increase the bar diameter or the number of rows. Otherwise the routine at lines 1410 to 1440 finds the smallest area in excess of A1 and enters the row location of this element into J1. Thus the content of A(1,J1) represents the required centres.

2.3.4.9 Block (L): Statements 1450–1670

Block (L) has the same function in this program as was described in Section 2.2.4.10. The steel percentage (P1), which in conjunction with the specified slab depth satisfied the bending criteria, is used to determine an effective depth (D5) which meets deflection requirements. D5 is compared with the available effective depth D1 at line 1670. If the depth required for deflection is greater than that available then a jump to line 2060 instructs the designer to increase the slab depth.

2.3.4.10 Block (M): Statements 1680–2020

The construction and function of this block is identical to that described in Section 2.2.4.12. The actual shear stress (V4) and the allowable value (R9*V9) are compared at line 2020. If at this juncture the actual shear stress is less than the allowable value then the slab will have satisfied all the programmed design criteria. Therefore a jump to line 2170 informs the designer that the proposed slab constitutes a possible solution.

2.3.4.11 Block (N): Statements 2030–2040

The calculation will only enter this block if shear requirements dictate that the slab should be modified in some way. The concern of the program is to determine the nature of the modification. This depends upon the level of shear stress. If at line 2030 it is found that the actual shear stress (V4) is less than the absolute maximum permitted value (V8), then a jump to line 2090 indicates that a solution is possible if shear reinforcement is included. Otherwise a jump to line 2060 advises the designer to increase the slab depth.

2.3.4.12 Block (Q): Statements 2050–2450

There are four points of entry into this final block. At two of them (lines 2060 and 2140) the designer is advised to modify the slab section parameters and the design procedure automatically returns to a request for further input at line 870. The two remaining entry points (at lines 2100 and 2170) advise the designer that a solution is possible, although at his discretion the output of results may be suppressed (see line 2190) if he suspects that a better solution is available. If such a facility were not built into the program, then the results for any slab which met CP 110 requirements would automatically be output, a time wasting process if it is unlikely that the results would be accepted. Even so, if the designer elects for an output of results he may still choose between accepting them or rejecting them in favour of a further trial (see line 2360).

If at line 2360 the designer does accept the current solution, then the design procedure will return for a new problem data input at line 780. The program run can be terminated at this point by an input of five zeros, separated by commas.

2.3.4.13 Example of Program RC9 Output

A typical solution given by this program appears below. Following the input of basic design parameters the first design proposal was an overall slab depth of 250 mm together with one layer of 20 mm diameter bars. Whilst this combination showed that a solution was possible an output of results was rejected at this stage in favour of another trial. The modified slab depth of 200 mm proved to be too shallow. When this was increased to 220 mm a solution was again possible and an output of results was requested. This solution was accepted.

```
RC9

CONCRETE GRADE (FCU) TO BE CHOSEN FROM THE FOLLOWING
20,25,30,40 AND 50 N/MM↑2

STEEL GRADE (FY) TO BE CHOSEN FROM THE FOLLOWING
250,410,425,460 AND 500 N/MM↑2

SPAN(M),SUPERLOAD(KN/M↑2),CONCRETE WEIGHT(KN/M↑3),
FY(N/MM↑2), FCU(N/MM↑2)
 ? 5.2,5,24,250,20
SLAB DEPTH(MM),BAR DIAMETER(MM),NO. OF ROWS
 ? 250,20,1
SOLUTION POSSIBLE

IF RESULT REQUIRED TYPE 1 ELSE 0
 ? 0

SLAB DEPTH(MM),BAR DIAMETER(MM),NO. OF ROWS
 ? 200,20,1
INCREASE SLAB DEPTH

SLAB DEPTH(MM),BAR DIAMETER(MM),NO. OF ROWS
 ? 220,20,1
SOLUTION POSSIBLE

IF RESULT REQUIRED TYPE 1 ELSE 0
 ? 1

SLAB DEPTH= 220 MM
NUMBER OF LAYERS OF REINFORCEMENT= 1
BAR DIAMETER= 20 MM
```

```
BAR CENTRES= 200 MM
COVER TO MAIN REINFORCEMENT= 25 MM
CALCULATED STEEL AREA= 1447.47 MM↑2
STEEL AREA PROVIDED= 1571. MM↑2


IF SOLUTION IS ACCEPTABLE TYPE 1 ELSE 0 ? 1


NEXT PROBLEM

SPAN(M),SUPERLOAD(KN/M↑2),CONCRETE WEIGHT(KN/M↑3),
FY(N/MM↑2),FCU(N/MM↑2)
  ? 0,0,0,0,0


RUNNING TIME:    3.9 SECS   I/O TIME :    5.2 SECS
```

2.4. The Yield Line Analysis of Rectangular Slabs

2.4.1 Introduction

The factor which most constrains Program RC9 to dealing with simply supported, one-way spanning slabs is the monolithic adoption, at line 980, of a design moment coefficient equal to 1/8. But if it were arranged for the moment coefficient to take a variety of values then the program's range of application would be substantially improved.

By adopting the simply supported slab design moment in the first instance, this implied that collapse would follow on the formation of a linear plastic hinge (a yield line) at midspan and across the full width of the slab. Since the actual mode of collapse was obvious from the nature of the loading and the support conditions, no further investigation was necessary. But this is not so in the case of rectangular panels supported on all four edges where, even though the general form of the collapse mode is known, the actual mechanism must be established each time a new solution is attempted. Program YL4 was written to show how this kind of problem may be solved. In Section 2.4.4 methods of incorporating the program (or its results) with the parent design program are discussed.

2.4.2 Outline of Yield Line Analysis Theory

In this discussion the dimensions, hinge rotation angles and bending moments etc. are called by the names of the variables assigned to them in the accompanying Program YL4. And when the derived expressions can be related to specific lines in the program, this is noted.

The yield line analysis of a rectangular panel, which is supported on all four sides and carries a uniformly distributed collapse load, is executed in two stages. During the first stage in the analysis the collapse mechanism is assumed

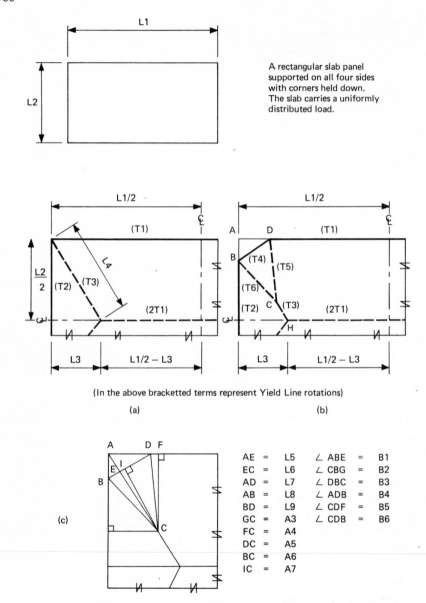

Figure 2.4 Rectangular panel collapse mechanisms

to take the form shown in Fig. 2.4a, where solid lines represent hogging moment yield lines and chained lines represent sagging moment yield lines.

In this first stage the ultimate value of the short span sagging moment is called M2; but for the purpose of deriving the internal work component of the work equation, M2 is taken as unity. It is assumed that the slab yield line moments are related in the following way:

the short span sagging moment/hogging moment ratio = M1   line 310
the long span sagging moment/hogging moment ratio = M8   line 310
and the ratio of long span/short span sagging moment = M9   line 310

It follows from this and Fig. 2.4a that the moment of resistance along the diagonal yield line will therefore be:

$$M7 = (L3/L4)^2 + M9*(L2/2*L4)^2 \hspace{3cm} \text{line 450}$$

If, as the slab begins to collapse, the long support hinges are assumed to rotate through unit angle, then:

the rotation of the hinge at midspan, parallel to the long side = 2, the rotation of the hinge along the short span side is

$$T2 = L2/(2*L3) \hspace{5cm} \text{line 430}$$

and the rotation of the hinge along the diagonal yield line is

$$T3 = L3/L4 + L2^2/(4*L3*L4) \hspace{3.5cm} \text{line 440}$$

From the pattern of yield lines it is seen that the collapse mechanism is one in which triangular areas of slab rotate about the short sides and trapezoidal areas rotate about the long sides. The external work done by a unit uniformly distributed collapse load will therefore be:

$$W2 = (T2*L2*L3^2)/3 + (L3*L2^2)/6 + (L1 - 2*L3)*L2^2/4 \hspace{1cm} \text{line 470}$$

And the internal work done by the rotating hinges will be:

$$\begin{aligned} W1 = {} & 2*L1/M1 + (2*L2*T9*M9)/M8 + (L1 - 2*L3)*2 \\ & + (4*L4 - T3*M7) \hspace{4cm} \text{line 460} \end{aligned}$$

If the short span sagging moment hinge is in fact worth M2/unit length, then the internal work done becomes W1*M2, and by equating internal and external work we get that:

$$M2 = W2/W1 \hspace{5cm} \text{line 480}$$

The geometry of the Fig. 2.4a mechanism is completely defined when the dimension L3 is known. The correct value of L3 is the one for which M2 is at a maximum. In the program this result is established through a systematic trial and error procedure.

At this stage we have only found the correct form of a *possible* mechanism, one which is not the actual mode of collapse although it is often accepted as such even though it predicts plastic moments up to 10% less than those actually required. This lower value is usually accepted on the argument that the reserve of strength offered by membrane force actions (which are not taken account of in this analysis) more than compensates for the marginal lack of predicted strength in pure bending. Fig. 2.4b shows a more likely collapse mode—one in which triangular segments are assumed to form along what were previously single diagonal yield lines. The analysis of this mode comprises the second stage in the operation. The primary reason for undertaking the first analysis was to establish an accurate location for the point at which the pair of diagonal yield lines and the midspan yield line intersected. In the present analysis what was previously assumed to be a single diagonal yield line now splits into two branches somewhere along its length, both of which intersect the edges of the panel. If the corners are held down a hogging yield line across each corner completes the mechanism.

The second stage analysis follows the course established by the first one in that an equation relating internal work to external work must again be established. In this it is marginally simpler to modify the existing work equation than to derive another from first principles.

Referring to Figs. 2.4b and c, the hinge rotations along the yield lines defining the new triangular segments are:

$$T4 = A4/A7 \qquad\qquad \text{line 750}$$
$$T5 = T4*\text{Cos } B6 + \text{Cos } B5 \qquad\qquad \text{line 760}$$
$$T6 = T4*\text{Cos } B3 + T2*\text{Cos } B2 \qquad\qquad \text{line 770}$$

And the ultimate moments of resistance along these yield lines are:

$$M0 = L7^2/(L9^2*M1) + (L8^2*M9)/(L9^2*M8) \qquad\qquad \text{line 800}$$
$$M5 = (A3 - L7)^2/A5^2 + M9*A4^2/A5^2 \qquad\qquad \text{line 780}$$
$$M6 = M9*(A4 - L8)^2/A6^2 + A3^2/A6^2 \qquad\qquad \text{line 790}$$

The original value of W2 must be modified to take account of:

1. W6—the additional work done by the load carried by the triangular area BCD; and
2. W4—the work component represented by the load carried by the triangular areas ABC and ACD, already taken into account and now to be subtracted.

Where $W6 = 2*L9*A7^2*T4/3$ \qquad\qquad line 820
and $W4 = 2*L8*A3^2*T2/3 + 2*L7*A4^2/3$ \qquad\qquad line 650

Hence the external work done by the unit collapse load is:

$$W9 = W2 - W4 + W6 \qquad\qquad \text{line 840}$$

The original value of W1 must be modified to take account of:

1. W5—the additional work done by the rotating hinges BC, CD and BD; and
2. W3—the internal work done at the yield lines AB, AC and AD, already taken into account and now to be subtracted.

Where $W5 = 4*L9*T4*M0 + 4*A5*T5*M5 + 4*A6*T6*M6$ line 810
and $W3 = 4*L7/M1 + 4*L8*T2*M9/M8 + 4*(L5 + L6)*T3*M7$ line 640

Hence the internal work done by the rotating hinges is:

$$W8 = W1 - W3 + W5 \hspace{4cm} \text{line 830}$$

If in this analysis the short span sagging moment of resistance has a value of M4/unit length, then the internal work done becomes $W8*M4$, and on equating external and internal work we get:

$$M4 = W9/W8 \hspace{4cm} \text{line 850}$$

The proportions of the triangular segment BCD (see Fig. 2.4c) affects the value of M4. The combination of the parameters L6, L7 and L8 which gives a maximum value of M4 is established in the program by means of a trial and error procedure.

2.4.3 Program Specification—YL4

The purpose of this program is to demonstrate the computer aided analysis of rectangular slabs by the *Yield Line* method. If, as is the case with this program, the analysis is restricted to a single panel type and loading condition, then the procedure can be an automatic one. Program YL4 analyses rectangular slabs supported on all four sides, and it further assumes that the corners are held down. The single loading condition considered is that of a uniformly distributed load covering the whole area of the panel.

Whilst the analysis assumes that equal moments are generated at pairs of opposite sides, the moments in the long and short span directions need not necessarily be the same, and neither need the individual span/support moment ratios. An important restriction to the application of this program is that it is assumed that the supporting beams have sufficient strength to allow the slab to develop the yield line pattern on which the analysis is based.

The program output gives the coordinates of the points at which the assumed yield lines intersect together with a list of design moment coefficients related to a collapse load of W/unit area.

2.4.4 Program Description—YL4

A list of the variables used in the program is given on the next page, and followed by the program listing.

64

A3, A4, A5, A6, A7 — Lengths defined in Fig. 2.4c

B1, B2, B3, B4, B5, B6 — Angles defined in Fig. 2.4c

D1 — Increment of L3

L1, L2, L3, L4 — Lengths defined in Fig. 2.4a

L5, L6, L7, L8, L9 — Lengths defined in Fig. 2.4c

M0 — Plastic moment at BD

M1 — Moment ratio (A)

M2 — Short span plastic moment in 1st analysis

M4 — Short span plastic moment in 2nd analysis

M5 — Plastic moment at DC

M6 — Plastic moment at BC

M7 — Plastic moment at AH

M8 — Moment ratio (B)

M9 — Moment ratio (C)

Z4 — Suggested long/short span sagging moment ratio

Z5 — Percentage excess of M4 over M2

T2, T3, T4, T5, T6 — Yield line rotations defined in Figs. 2.4a and b

W1, W2 — Terms in 1st work equation

W3, W4, W5, W6 — Terms modifying 1st work equation

W8, W9 — Terms in 2nd work equation

Z1, Z2, Z3, Z8, Z9 — Contents define the final mechanism

Z6, Z7 — Switches

```
100 PRINT "INPUT PANEL DIMENSIONS *********:*"
110 PRINT
120 PRINT "LONG SPAN AND SHORT SPAN (M)";
130 INPUT L1,L2
140 PRINT
150 IF L1/L2<3 THEN 180
160 Z4=0.3636
170 GOTO 190
180 Z4=1/(2.75-(3-L1/L2)*2/4*1.75)
190 IF Z6=1 THEN 270
200 PRINT "SHORT SPAN - SAGGING/HOGGING MOMENT RATIO    (A)"
210 PRINT " LONG SPAN - SAGGING/HOGGING MOMENT RATIO    (B)"
220 PRINT "LONG SPAN/SHORT SPAN SAGGING MOMENT RATIO    (0)"
230 PRINT
240 PRINT "ACCEPTABLE RANGE OF (A) AND (B) IS 0.6667 TO 1.0"
250 PRINT
260 PRINT "ACCEPTABLE RANGE OF (C) IS 0.3636 TO 1.0"
270 PRINT
280 PRINT "SUGGESTED RATIOS ARE - (A)=0.75, (B)=0.75 AND (C)="Z4
290 PRINT
300 PRINT "INPUT CHOSEN VALUES OF (A),(B) AND (C)"
310 INPUT M1,M8,M9
320 PRINT
330 PRINT
340 L3=0
350 Z1=0
360 D1=0.05*L2
370 GOTO 390
380 Z1=M2
390 L3=L3+D1
400 IF L3<=L1/2 THEN 420
410 L3=L1/2
420 L4=SQR(L3*2+(L2*2)/4)
```

```
430  T2=L2/(2*L3)
440  T3=L3/L4+(L2+2)/(4*L3*L4)
450  M7=L3+2/L4+2+M9*L2+2/(4*L4+2)
460  W1=2*L1/M1+2*L2*T2*M9/M8+(L1-2*L3)*2+4*L4*T3*M7
470  W2=(T2*L2*L3+2)/3+(L3*L2+2)/6+(L1-2*L3)*L2+2/4
480  M2=W2/W1
490  IF L3=L1/2 THEN 540
500  IF M2>Z1 THEN 380
510  D1=-D1/2
520  IF ABS(D1)<0.02 THEN 540
530  GOTO 380
540  Z1=0
550  FOR L8=0.1*L2 TO 0.35*L2 STEP 0.05*L2
560  FOR L7=0.1*L2 TO 0.35*L2 STEP 0.05*L2
590  L5=SQR(A1+2+A2+2)
600  FOR L6=0.8*(L4-L5) TO (L4-L5) STEP 0.05*(L4-L5)
610  A3=(L5+L6)*L3/L4
620  A4=(L5+L6)*L2/(2*L4)
630  L9=SQR(L8+2+L7+2)
640  W3=4*L7/M1+4*L8*T2*M9/M8+4*(L5+L6)*T3*M7
650  W4=2*L8*A3+2*T2/3+2*L7*A4+2/3
660  A6=SQR((A4-L8)+2+A3+2)
670  A5=SQR((A3-L7)+2+A4+2)
680  B1=ATN(L7/L8)
690  B2=ATN(A3/(A4-L8))
700  B3=3.1416-B1-B2
710  B4=ATN(L8/L7)
720  B5=ATN(A4/(A3-L7))
730  B6=3.1416-B4-B5
740  A7=A6*SIN(B3)
750  T4=A4/A7
760  T5=T4*COS(B6)+COS(B5)
770  T6=T4*COS(B3)+T2*COS(B2)
780  M5=(A3-L7)+2/A5+2+M9*A4+2/A5+2
790  M6=M9*(A4-L8)+2/A6+2+A3+2/A6+2
800  M0=L7+2/(L9+2*M1)+L8+2*M9/(L9+2*M8)
810  W5=4*L9*T4*M0+4*A5*T5*M5+4*A6*T6*M6
820  W6=2*L9*A7+2*T4/3
830  W8=W1-W3+W5
840  W9=W2-W4+W6
850  M4=W9/W8
860  IF Z1>M4 THEN 920
870  Z1=M4
880  Z2=A3
890  Z3=A4
900  Z8=L7
910  Z9=L8
920  NEXT L6
930  NEXT L7
940  NEXT L8
950  PRINT "IF INFORMATION IS REQUIRED CONCERNING THE"
960  PRINT "YIELD LINE PATTERN THEN TYPE 1 ELSE 0";
970  INPUT Z7
980  Z5=(Z1-M2)*100/M2
990  IF Z7=0 THEN 1140
1000 PRINT
1010 PRINT "YIELD LINE PATTERN HAS CORNER SEGMENTS"
1020 PRINT
1030 PRINT
1040 PRINT "MEASURED FROM THE CENTRE OF THE PANEL"
1050 PRINT "THE COORDINATES (M) OF THE POINTS OF"
1060 PRINT "INTERSECTION OF THE YIELD LINES IN ONE"
1070 PRINT "QUARTER OF THE PANEL ARE :"
1080 PRINT
1090 PRINT "    X              Y"
1100 PRINT (L1/2-L3)," 0"
```

```
1110 PRINT (L1/2-Z2),(L2/2-Z3)
1120 PRINT (L1/2-Z8),L2/2
1130 PRINT L1/2,(L2/2-Z9)
1140 PRINT
1150 PRINT
1160 PRINT "REQUIRED SHORT SPAN SAGGING MOMENT="Z1"*W KNM/M"
1170 PRINT "REQUIRED SHORT SPAN HOGGING MOMENT="Z1/M1"*W KNM/M"
1180 PRINT " REQUIRED LONG SPAN SAGGING MOMENT="Z1*M9"*W KNM/M"
1190 PRINT " REQUIRED LONG SPAN HOGGING MOMENT="Z1*M9/M8"*W KNM/M"
1200 IF Z7=0 THEN 1240
1210 PRINT
1220 PRINT "THE REQUIRED MOMENTS ARE"Z5"% GREATER THAN THOSE"
1230 PRINT "PREDICTED BY THE BASIC Y-YIELD LINE PATTERN"
1240 PRINT
1250 PRINT
1260 PRINT
1270 PRINT
1280 PRINT "IF ANOTHER PANEL IS TO BE ANALYSED THEN TYPE 1 ELSE 0";
1290 INPUT Z6
1300 IF Z6=1 THEN 100
1310 END
```

2.4.4.1 Block (A): Statements 100–330

Two sets of design parameters are input at this block; these are the panel dimensions at lines 100 to 130 and the moment ratios at lines 300 to 310. Even though the assessment of the moment ratios (A), (B), and (C)—see lines 200 to 220—is largely a matter for individual judgement, the program user will benefit from being offered guidance when choosing these values, particularly in the case of ratio (C) where no firm value can be assigned unless the slab is assumed to be isotropic, in which case the ratio of moments will be 1.0.

In a two-way spanning slab the proportion of the load carried by the long span is a function of the long span/short span ratio l_y/l_x. For $l_y/l_x = 1.0$ an equal amount of load is carried in each direction and the long span claims a smaller proportion of the total load as l_y increases. It is usually accepted that for $l_y/l_x \geq 3.0$ the middle strip width of $(l_y/l_x - 1.0)*l_x$ tends to shed the whole of its load in the short span direction whilst the outer strips adjacent to the short sides continue to distribute their load in two directions. This argument is substantially supported by the evidence of actual yield line patterns at collapse. In order to reflect this behaviour in yield line design (the term 'design' is used here to emphasize that the outcome of an analysis will be translated into steel and concrete) the procedure at lines 150 to 180 makes an assessment of the relative moments and prints out a recommended ratio at lines 260 to 280 for the designer to accept or reject according to his own preference. This assessment is based upon the information given in Fig. 2.5 where it is assumed that regardless of the l_y/l_x ratio the maximum long span free bending moment will be $Wl_x^2/22$ (i.e. the maximum free moment at collapse in a square panel of side l_x) and that the maximum short span moment will vary parabolically from $Wl_x^2/22$ to $Wl_x^2/8$ for $1.0 \leq l_y/l_x \leq 3.0$. The value of l_y/l_x (i.e. L1/L2) at line 150 determines the basis on which Z4, the ratio of moments, will be calculated. For $l_y/l_x > 3.0$, Z4 will become 0.3636 (i.e. $(Wl_x^2/22)/(Wl_x^2/8)$) at line 160. Otherwise Z4 is

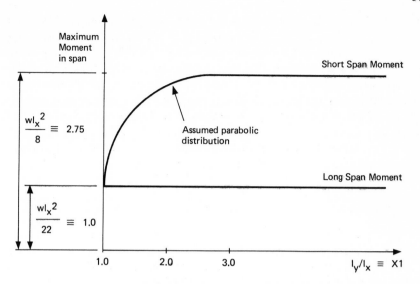

Figure 2.5 Assumed variation in span moment ratios

calculated according to the expression at line 180 which follows from Fig. 2.5.

Limiting the span moment/support moment ratio in each direction to the range recommended in CP 110 ensures that there will be no undue cracking under service load conditions. The program user is reminded of the acceptable range at line 240, and suggested ratios are given at line 280, before a decision must be made at line 300. The printout at lines 200 to 260 occurs only once per program run; since subsequent solutions require that $Z6 = 1$ the program jumps from line 190 to line 270 when $Z6$ has this value.

2.4.4.2 Block (B): Statements 340–530

The purpose of this block of statements is to determine the value of L3 (see Fig. 2.4a) for which the plastic moment of resistance M2 has a maximum value. Neglecting at this stage the way in which trial values are assigned to L3, for a given value of L3 M2 is calculated at lines 420 to 480. At line 500 the current value of M2 is compared with the content of Z1, a variable which stores the previous value of M2 (see lines 500,530 and 380). In effect, at the end of each trial, a check is made upon the slope of the (L3, M2) function. If M2 is greater than Z1 then the peak value has not yet been attained; if M2 is less than Z1 then the peak value has been passed. The current value of M2 compared with its previous value therefore indicates whether the next trial value of L3 should be increased or decreased.

L3 is assigned trial values at line 390 where each new value is the algebraic sum of the previous value of L3 and an increment called D1. D1 is set initially to an arbitrary value of $0.05*L2$ at line 360 and thereafter it retains this value until the check at line 500 indicates that the peak value of M2 has been passed. When this happens the increment is halved at line 510 and its sign is changed. Further trial values of L3 therefore become successively smaller until the peak value is passed once more, this time on the return. This results in D1 being halved again, and there is an accompanying change of sign. In this way trial values of L3 oscillate about the true value, a process which continues until the increment D1 is small enough to neglect. This fact is recognized at line 520 where an arbitrary lower limit of $\pm\ 0.02$ is given to the value of D1. A limitation that L3 can never be greater than L1/2 is imposed at line 410, but this only has cause to operate in the case of square isotropic panels.

2.4.4.3 Block (C): Statements 540–940

The form of the first mechanism is now modified to include a triangular segment at each corner; but at the beginning of this stage the degree to which the original diagonal yield lines are replaced is unknown. The new segment is defined by the dimensions L6, L7 and L8; FOR statements at lines 550, 560 and 600 limit the range of these dimensions over which a solution is sought. The limits of the envelope defined by these FOR statements were determined by using the program itself to solve boundary problems to indicate the probable upper and lower limits to the dimensions of the triangular segment.

Another factor which affects the number of trial solutions, and hence the computing time required, is the size of the step incorporated in the FOR statements. This was resolved by evaluating the results of further program runs in which the step size was varied. It was found that the relatively coarse increments which were finally adopted give plastic moments within 0.1% of those predicted when much finer steps are used.

For each set of trial dimensions the calculation follows the course outlined in Section 2.4.2 which culminates at line 850 with an expression for M4, the required plastic moment of resistance in the short span direction. The variable called Z1, which is initially set to zero at line 540, subsequently takes (at line 870) the largest value of M4 attained at any stage in the calculation—see the statement at line 860 where a jump is made to the next set of trial values if at that stage Z1 is greater than the current value of M4. The dimensions which define the mechanism requiring the largest value of M4 are stored at lines 880 to 910.

2.4.4.4 Block (D): Statements 950–1310

This block is concerned with the output of information. If the form of the yield line pattern is of no interest then this information may be suppressed by setting the variable Z7 to zero at line 970. The opportunity to carry out further analyses during the same program run is given at lines 1280 to 1300 if Z6 is set to 1.

2.4.5 The Conversion of RC9 into a Two-way Spanning Slab Program

Before considering the program modifications in detail, a policy decision had to be made regarding the way in which the slab moment coefficients were to be incorporated with the parent design program. There are three ways of doing this:

1. To bring the analysis program YL4 and a modified version of RC9 together in one program (a simple matter of deleting the END statement in YL4 and renumbering the lines in RC9 so that no overlapping occurs). In this case the moment coefficients are generated within the overall design program and no double handling of information is necessary.
2. To employ a version of YL4 to produce, in a one-off operation, the limit analysis equivalent of the Table 13 CP 110 slab bending moment coefficients. The new array of coefficients would then be permanently incorporated in a modified version of RC9.
3. To run the program YL4 and a modified version of RC9 independently of each other (and in that order) each time a problem is to be solved and to use the moment coefficient output of YL4 as part of the input to RC9. This method anticipates the 'Library' approach to decision design which is discussed in Chapter 3.

In the version presented here option 3 was chosen, primarily because it required fewer changes to be made to the existing Program RC9.

A list of the statements needed to convert Program RC9 into one which is capable of processing two-way spanning slab problems is given below. When these statements are superimposed upon RC9 the resultant program is called TWS1.

The statements fall into six categories:

1. Those that are necessary to introduce the two-way spanning slab concept (some of which are additional to RC9 whilst others replace existing statements);
2. New statements which reroute the basic slab design calculation according to which moment coefficient is currently being considered;
3. New statements which introduce the continuous slab deflection criteria;
4. Existing RC9 statements which are modified to take account of the single layer of reinforcement in either direction at a given face;
5. Existing jump statements now modified to suit the new program structure;
6. Existing RC9 statements which now have no role in the new program (signified by line numbers which appear on their own).

The four moment coefficients are the first design parameters to be input to Program TWS1—see lines 791 to 794. They are read into the first row of an array called B( ) in the order dictated by the statements at lines 792 and 793. Whenever they are assigned zero values then an exit is made from the program at line 795.

Statements needed to convert Program RC9 to Program TWS1

```
791  PRINT "INPUT DESIGN MOMENT COEFFICIENTS IN THE ORDER ******"
792  PRINT "SHORT SPAN - SAGGING, HOGGING"
793  PRINT " LONG SPAN - SAGGING, HOGGING"
794  INPUT B(1,1),B(1,2),B(1,3),B(1,4)
795  IF B(1,1)=0 THEN 2450
800  PRINT "SHORT SPAN (M), SUPERLOAD (KN/M↑2),"
801  PRINT "CONCRETE WEIGHT (KN/M↑3)"
830  IF L1<=20 THEN 870
860
870  PRINT "SLAB DEPTH (MM),BAR DIAMETER (MM)"
880  INPUT T9,B9
882  Z=0
891  Z=Z+1
892  IF Z=5 THEN 2170
893  IF Z=1 THEN 900
894  IF Z>2 THEN 897
895  D2=B(3,1)
896  GOTO 970
897  D2=B(3,1)+B9
898  GOTO 970
960  D2=B9/2+C2
961  B(3,1)=D2
980  M1=(1.6*W1+1.4*T9*W3/1000)*B(1,Z)
1100 P1=A1/(10*D1)
1110
1442 B(2,Z)=A(1,J1)
1443 IF Z>1 THEN 891
1625 IF B(1,2)>0 THEN 1645
1645 D1=26
1646 GOTO 1660
1650 IF B(1,2)>0 THEN 1657
1655 D1=30-L1
1656 GOTO 1660
1657 D1=36-L1
2020 IF V4<=R9*V9 THEN 891
2100 PRINT "IF SOLUTION IS POSSIBLE THEN SHEAR"
2101 PRINT "REINFORCEMENT IS REQUIRED."
2120 GOTO 891
2140 PRINT "INCREASE BAR DIAMETER"
2250 PRINT "COVER TO SHORT SPAN REINFORCEMENT="C2"MM"
2260 PRINT
2270 PRINT "SHORT SPAN REINFORCEMENT **********"
2280 PRINT "   SPAN STEEL :"B9"MM DIAMETER AT"B(2,1)"MM CRS"
2285 IF B(1,3)=0 THEN 2340
2290 PRINT "SUPPORT STEEL :"B9"MM DIAMETER AT"B(2,2)"MM CRS"
2300 PRINT
2310 PRINT " LONG SPAN REINFORCEMENT **********"
2320 PRINT "   SPAN STEEL :"B9"MM DIAMETER AT"B(2,3)"MM CRS"
2330 PRINT "SUPPORT STEEL :"B9"MM DIAMETER AT"B(2,4)"MM CRS"
```

The long span dimension is of no interest in this calculation; it was taken into account when the moment coefficients were determined by Program YL4. But since the deflection and shear criteria are a function of the short span this dimension is still needed and is input at line 880. Note also at this line that since there is no call for the number of rows of reinforcement it is assumed that there will be only one row in each direction at a given face.

The basic blocks of program in TWS1 are essentially the same as those of RC9 (see Section 2.3.2) and although it processes the four panel design moments automatically the program still functions in a decision design capacity.

In TWS1 line 891 begins Block (F). A variable called Z, which is set initially to zero at line 882, is advanced by 1 at line 891 as each moment coefficient is called in turn. The value of Z serves to locate the position of the moment coefficient currently being considered (i.e. $B(1,Z)$) and also influences the program route at each stage.

$$Z = 2, 3 \text{ or } 4$$

Thus: $\quad A \rightarrow B \rightarrow C \rightarrow F \rightarrow V \rightarrow W \rightarrow X \rightarrow G \rightarrow E \rightarrow L \rightarrow M \rightarrow N \quad Q$

$$Z = 1$$

$$Z = 5$$

When $Z = 1$ Blocks (F) to (N) inclusive play a role in the design calculation and the initial path therefore follows that of program RC9. At this stage the short span sagging moment coefficient $B(1,1)$ determines the moment against which the proposed slab is checked and the reinforcement is designed. During this calculation the value of D2 (i.e. the distance from the tension face to the centre of gravity of the reinforcement) is stored in $B(3,1)$, at line 961, for later reference. The additional statements at Block (F)—see lines 1625 and 1645 to 1657—take into account the possibility that the short span may be continuous. In this case an alternative deflection criterion is imposed. A failure of the proposed slab at any stage to meet the programmed design criteria results in a jump to the beginning of Block (Q) where a deeper slab or an increased bar diameter is advised. If the proposed slab survives the first stage then a return to Block (F) advances the value of Z.

For $Z = 2,3$ or 4, since it is now only necessary to consider bending criteria, the design procedure is confined to a loop which embraces Blocks (F) to (E). The statements at lines 894 to 898 control the current value of D2 and hence the value of D1 on which the design for bending is based. D2 has the same value whether Z is equal to 1 or 2 (see lines 894 to 896). But for $Z > 2$, since the long span reinforcement is assumed to form a second layer, the value of D2 is increased by one bar diameter (see lines 894 and 897). If Z succeeds in reaching the value of 5 then the proposed slab will have fulfilled all the programmed requirements and a jump from line 892 to Block (Q) allows for the possibility of an output of results.

2.4.6 Examples of Output from Programs YL4 and TWS1

Typical solutions given by these programs appear below. These solutions are considered together because they are concerned with different aspects of the same problem, i.e. the design of a rectangular slab, 6.75 m × 4.25 m, which is supported continuously on all four sides.

The slab is analysed by Program YL4. Following the input of panel dimensions the designer chose the same moment ratios (A), (B), and (C) as were suggested by the program output. When the solution was available he elected for a full output of results which included information concerning the yield line pattern in addition to the moment coefficients.

The moment coefficient output from YL4 comprised the first input to Program TWS1. Thereafter the running of this program was identical to that of RC9 (see Section 2.3.4.13). In the present instance four attempts were made before the designer elected to accept a solution.

```
YL4

INPUT PANEL DIMENSIONS ************

LONG SPAN AND SHORT SPAN (M) ? 6.75,4.25

SHORT SPAN - SAGGING/HOGGING MOMENT RATIO     (A)
 LONG SPAN - SAGGING/HOGGING MOMENT RATIO     (B)
LONG SPAN/SHORT SPAN SAGGING MOMENT RATIO     (C)

ACCEPTABLE RANGE OF (A) AND (B) IS 0.6667 TO 1.0

ACCEPTABLE RANGE OF (C) IS 0.3636 TO 1.0

SUGGESTED RATIOS ARE - (A)=0.75, (B)=0.75 AND (C)= .532474

INPUT CHOSEN VALUES OF (A),(B) AND (C)
 ? 0.75,0.75,0.53


IF INFORMATION IS REQUIRED CONCERNING THE
YIELD LINE PATTERN THEN TYPE 1 ELSE 0 ? 1

YIELD LINE PATTERN HAS CORNER SEGMENTS


MEASURED FROM THE CENTRE OF THE PANEL
THE COORDINATES (M) OF THE POINTS OF
INTERSECTION OF THE YIELD LINES IN ONE
QUARTER OF THE PANEL ARE :


      X                 Y
   1.32969           0
   1.32969           2.91038E-11
   2.7375            2.125
   3.375             1.275


REQUIRED SHORT SPAN SAGGING MOMENT= .616014 *W KNM/M
REQUIRED SHORT SPAN HOGGING MOMENT= .821352 *W KNM/M
 REQUIRED LONG SPAN SAGGING MOMENT= .326488 *W KNM/M
 REQUIRED LONG SPAN HOGGING MOMENT= .435317 *W KNM/M


THE REQUIRED MOMENTS ARE 7.43672 % GREATER THAN THOSE
PREDICTED BY THE BASIC Y-YIELD LINE PATTERN
```

IF ANOTHER PANEL IS TO BE ANALYSED THEN TYPE 1 ELSE 0 ? 0

RUNNING TIME: 5.3 SECS   I/O TIME :   2.6 SECS

TWS 1

CONCRETE GRADE (FCU) TO BE CHOSEN FROM THE FOLLOWING
20,25,30,40 AND 50 N/MM↑2

STEEL GRADE (FY) TO BE CHOSEN FROM THE FOLLOWING
250,410,425,460 AND 500 N/MM↑2

INPUT DESIGN MOMENT COEFFICIENTS IN THE ORDER **********
SHORT SPAN - SAGGING, HOGGING
 LONG SPAN - SAGGING, HOGGING
 ? 0.616,0.821,0.326,0.435
SHORT SPAN (M), SUPERLOAD (KN/M↑2),
CONCRETE WEIGHT (KN/M↑3), FY (N/MM↑2), FCU (N/MM↑2)
 ? 4.25,10.5,24,250,20
SLAB DEPTH (MM),BAR DIAMETER (MM)
 ? 140,12
SOLUTION POSSIBLE

IF RESULT REQUIRED TYPE 1 ELSE 0
 ? 0

SLAB DEPTH (MM),BAR DIAMETER (MM)
 ? 100,12
INCREASE SLAB DEPTH

SLAB DEPTH (MM),BAR DIAMETER (MM)
 ? 130,12
SOLUTION PO SSIBLE

IF RESULT REQUIRED TYPE 1 ELSE 0
 ? 1

SLAB DEPTH= 130 MM
COVER TO SHORT SPAN REINFORCEMENT= 25 MM

SHORT SPAN REINFORCEMENT **********
   SPAN STEEL : 12 MM DIAMETER AT 150 MM CRS
SUPPORT STEEL : 12 MM DIAMETER AT 100 MM CRS

 LONG SPAN REINFORCEMENT **********
   SPAN STEEL : 12 MM DIAMETER AT 250 MM CRS
SUPPORT STEEL : 12 MM DIAMETER AT 200 MM CRS

IF SOLUTION IS ACCEPTABLE TYPE 1 ELSE 0 ? 1

NEXT PROBLEM

INPUT DESIGN MOMENT COEFFICIENTS IN THE ORDER **********
SHORT SPAN - SAGGING, HOGGING
 LONG SPAN - SAGGING, HOGGING
 ? 0,0,0,0

RUNNING TIME: 4.3 SECS   I/O TIME :   6.3 SECS

Chapter 3

The Application of Decision Design Methods to the Solution of Prestressed Concrete Beam Problems

3.1 Introduction

Prestressed concrete design is an open-ended problem for which, in theory, there is an infinity of solutions. Whilst experience dictates that most of them be rejected, even within the constraints of a set of design regulations the threshold of acceptance or rejection will vary from one designer to another. This and the fact that a number of sections of different proportions could equally serve the same function demands an open-ended design approach which can rapidly produce several solutions for comparison. In the computer aided design context these needs are most readily met by applying decision design techniques.

A decision design program may take one of two basic forms. Either it is written as a single program which embraces the whole problem and the programmed calculation is made to halt at intervals for the input of design decisions, or the design problem is partitioned and a sub-program is written to process each individual stage. In the latter case the number of sub-programs comprising the 'library' would normally depend upon the number of design decisions which had to be taken along the route to a solution. These sub-programs would be run in a predetermined order, the output from one furnishing the information on which the designer made his next decision.

Both of these methods are discussed in Chapter 5 in relation to the problem of designing continuous reinforced concrete frames. The second method is discussed in this chapter with particular reference to the design of prestressed concrete beams for bending. Of the two methods the second is marginally the simplest to program. It also has the advantage of allowing the computer line to be freed for other operations whilst the next design step is contemplated.

Three basic design decisions are involved in the proportioning of a prestressed concrete section to meet service load bending requirements. These are:

1. To choose the section shape and its dimensions;
2. To select an eccentricity for the prestressing force;

3. To decide upon an arrangement of tendons.

A further requirement is that the ultimate load characteristics of the adopted section must be evaluated. Accordingly, a library of four sub-programs was written to handle these aspects of the overall design problem. The file names and basic roles of these sub-programs are as follows:

PB4　proportions the section to meet service load bending requirements;
PB2　selects the number and type of tendons to be included in the section;
PB3　calculates the spacing of the tendons;
PB5　determines the ultimate moment of resistance of the section.

Each sub-program will be discussed in detail later in this chapter, briefly the overall structure and operation of the library is as follows. PB4 works with data which is based upon a design decision regarding the section shape and its dimensions. It checks the proposed section against service load bending requirements and indicates either that a solution is possible or that the section dimensions must be modified. PB4 is therefore recycled as many times as is necessary in order to achieve what at this stage is an acceptable solution. The output from PB4 includes the limiting values of the eccentricity of the prestressing force and the distance of the section centre of gravity to its outer fibres—sufficient information to allow the designer to make an informed guess concerning the maximum eccentricity required to house the (as yet unknown number of) tendons within the section. PB2 works with data which includes the chosen eccentricity and information about the prestressing system to output the number and type of tendons required. At this stage the designer may conclude that it is not feasible to accommodate these tendons within the proposed section and he will therefore need to reassess the section and recycle the first two sub-programs. When a section has finally been adopted the designer decides upon the general pattern of the tendon group (i.e. the number of rows and the number of tendons in each row). Using this information as data PB3 calculates the row spacing which will give the required eccentricity. PB5 takes data concerning the section shape and distribution of steel to determine the ultimate moment of resistance and compare it with the minimum requirement. Even at this stage the section may need to be modified, in which case the whole procedure is recycled.

In this way a library of sub-programs may be used interactively to generate a design result. The total design of a prestressed concrete member requires additional considerations—an assessment of the losses in prestress, a delineation of the cable zone, shear and end-block design and an estimate of deflection. A survey of the decisions necessary to solve each of these problems would determine the number of sub-programs required to complete the design library.

3.2 Outline of Prestressed Concrete Composite Beam Bending Theory

At the expense of little extra programming effort the first sub-program (PB4) to be discussed in this chapter was developed to investigate bending aspects

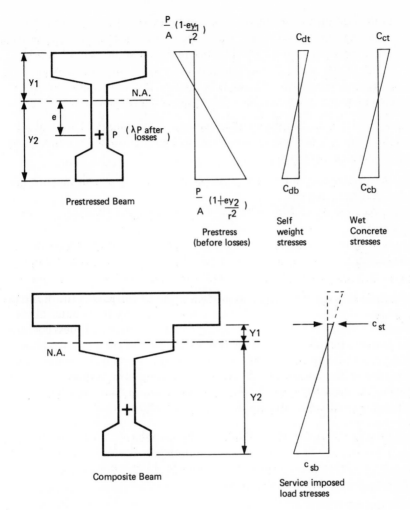

Figure 3.1 Normal stress distributions in prestressed and composite beams

of the *composite* prestressed concrete beam design problem. Even so, the program may equally be used to design normal prestressed concrete beam sections. This is a low level example of how the scope of a program may be enhanced by arranging for it to process a wider variety of problems.

In composite construction precast prestressed concrete beams and normal *in situ* concrete are brought together to form a section capable of sustaining the imposed service load. In the design of the prestressed beam component of the overall composite section it is usual to take account of bending stresses induced during the following three stages in construction:

1. The prestressed beams are cast and stressed before being lifted into their final position;

2. When in position the beams may be required either to support formwork, or to play the role of formwork, whilst the in situ concrete is placed (during this stage the beams still retain their original second moment of area);

3. After the *in situ* concrete has cured it becomes a monolithic part of the structure and subsequent loading is carried by the total composite section; the second moment of area is now that of the composite section.

Under service load conditions the assumed distributions of normal stress due to prestress and bending are as shown in Fig. 3.1 where the symbols used have the following meanings:

P is the prestressing force and γ the proportion of P which remains after losses have occurred;

e is the eccentricity of P, positive when measured below the centre of area of the section;

the cross-sectional area of the beam is A and r is its radius of gyration: the distances of the top and bottom extreme fibres of the prestressed beam from its centre of area are y_1 and y_2; those for the composite beam are Y_1 and Y_2.

The bending stresses at the top and bottom fibres of the prestressed beam are respectively:

c_{dt}, c_{db} due to the self-weight of the beam;
c_{ct}, c_{cb} due to the in situ concrete;
c_{st}, c_{sb} due to the imposed load acting on the composite section.
c and c_t are the allowable concrete stresses in compression and tension respectively. Compressive stresses are assumed to be positive.

The design of a section is accomplished by considering a set of inequations which relate the actual bending stress conditions at the extreme fibres to the allowable concrete stresses. These inequations take the following form:
considering the top fibres of the beam,

$$\frac{1}{P} \geq \frac{(ey_1/r^2 - 1)}{A(c_t + c_{dt})} \tag{3.1}$$

and either

$$\frac{1}{P} \geq \frac{(1 - ey_1/r^2)}{A(c - c_{dt} - c_{ct} - c_{st})} \qquad \text{(3.2a) if } c > c_{dt} + c_{ct} + c_{st}$$

or

$$\frac{1}{P} \leq \frac{\gamma\,(ey_1/r^2 - 1)}{A(c_{dt} + c_{ct} + c_{st} - c)} \qquad \text{(3.2b) if } c < c_{dt} + c_{ct} + c_{st}$$

considering the bottom fibres of the beam,

$$\frac{1}{P} \geq \frac{(1 + ey_2/r^2)}{A(c + c_{db})} \tag{3.3}$$

and

$$\frac{1}{P} \leq \frac{\gamma (1 + ey_2/r^2)}{A(c_{db} + c_{cb} + c_{sb} - c_t)} \tag{3.4}$$

When deriving inequations (3.2a), (3.2b) and (3.4) it was assumed that the combined stresses $(c_{ct} + c_{st})$ and $(c_{cb} + c_{sb})$ were greater than those resulting from loads imposed on the beam at the in situ concreting stage.

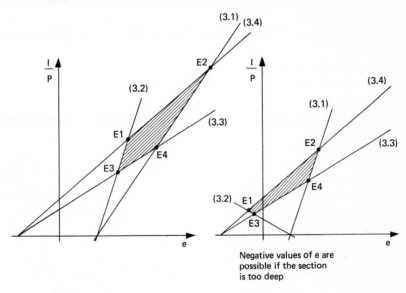

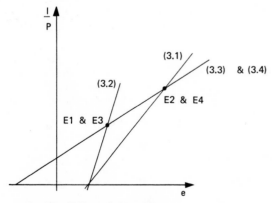

When lines (3.3) and 3.4) coincide
the minimum section depth is indicated.

Figure 3.2 Typical plots of $1/P$ v.e

By plotting the lines which represent the inequations satisfactory combinations of $1/P$ and e may be found from within the area bounded by the intersecting lines. Typical plots of $1/P$ v. e are given in Fig. 3.2, where the regions from which solutions can be taken are shown hatched.

3.3 Programming for the Decision Design of Composite Prestressed Concrete Beams to meet Bending Requirements

3.3.1 Program Specification—PB4

This sub-program checks the response of a designer specified composite prestressed concrete beam section to service load bending and compares this with the CP 110 recommendations. The beam is assumed to be simply supported and to carry a uniformly distributed imposed load.

3.3.2 The PB4 Flow Diagram

The flow diagram for this sub-program is shown in Fig. 3.3. Bending stresses in the prestressed component of the composite section must be determined both before and after the *in situ* concrete has been placed. Because the shape of the composite section may bear no resemblance to that of the original prestressed concrete beam it is necessary to input two sets of section dimensions. The form of the check design calculation makes it expedient to input the composite section dimensions at Block (B) followed by the prestressed beam dimensions at Block (F).

With reference to Fig. 3.3 the path taken by the design calculation is:

$$A \rightarrow B \rightarrow C \rightarrow D \rightarrow E \rightarrow F \rightarrow D \rightarrow I \rightarrow J \rightarrow K \rightarrow (G, H \text{ and } L) \text{ or } (M)$$

The parameters of span, loading, allowable stresses etc., are defined at Block (A). The sets of section dimensions input at Blocks (B) and (F) are duly processed; if they are able to yield a solution then the results are output at Block (L), if not then a 'section fail' message is printed out at Block (M). In either case the calculation returns to Block (B) where the program run may be terminated if desired.

3.3.3 Description of Sub-program PB4

A list of the variables and arrays used in this sub-program is given below.

A( )   Temporary store for   Q   Counter
       section dimensions   Q1   Switch

A1 Area of either section
A9 Area of composite section
B( ) Temporary store for section dimensions
B Width of asterisk block
B1⎫
B2⎪
B8⎬ Section dimensions
B9⎭
C( ) Temporary store for section dimensions
C1 c
C2 c_t
D Depth of asterisk block
D1⎫
D2⎭ Section dimensions
E1⎫
E2⎪ Allowable
E3⎬ eccentricities
E4⎭
F Trigger
G Temporary store
I Row counter
I1 Second moment of area
I2⎫
I3⎬ Intermediate steps in I1
I4⎭
L1 Span
L3 Loss factor
M Required ultimate moment of resistance
M1 Prestressed beam self-weight moment
M2 Moment due to in situ concrete
M3 Composite beam self-weight moment
N Counter
P Counter

Q2⎫
Q3⎭ Section dimensions
R1⎫
R3⎭ Radii of gyration
S1⎫
S2⎪ Stresses in prestressed
S3⎬ beam due to M1 and M2
S4⎭
S5⎫ Stresses in
S6⎬ composite beam
S7⎭ due to M3
T1⎫
T2⎪
T3⎪
T4⎬ Section dimensions
T5⎪
T8⎪
T9⎭
V Temporary store
W1 Weight of prestressed beam/unit length
W2 Imposed load/unit length
W4 Weight of composite section/unit length
W8 Weight of in situ concrete/unit length
X Depth of block represented by asterisks
X1⎫
X2⎬ Denominators of inequations
X3⎪ (3.1), (3.2a), (3.3) and (3.4)
X4⎭
Y Width of block represented by asterisks
Y1⎫ Distances to extreme fibres
Y2⎬ from section c. of g.
Y3⎭
Z Number of spaces
Z2 Denominator of inequation (3.2b)
Z9 Trigger

The program is listed below and should be read in conjunction with the flow diagram shown in Fig. 3.3.

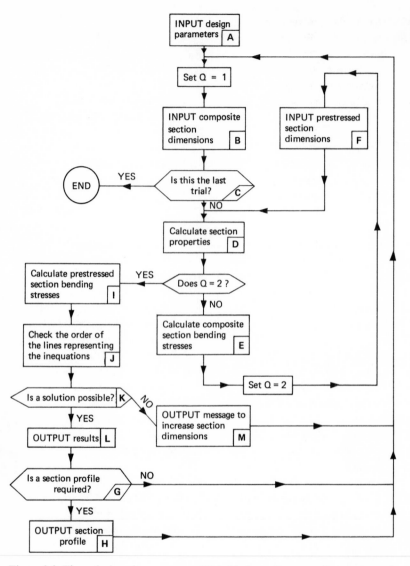

Figure 3.3 Flow design for program PB4 (design of composite prestressed concrete beam section to meet service load bending requirements)

```
100 PRINT "YOUR VALUES OF SPAN,SUPERLOAD,C,CT AND LOSSFACTOR ARE"
110 INPUT L1,W2,C1,C2,L3
120 LET Q1=0
130 LET Q1=Q1+1
140 IF Q1=2 THEN 200
150 PRINT
160 PRINT
170 PRINT "COMPOSITE SECTION DIMENSIONS ARE";
180 PRINT
190 IF Q1=1 THEN 240
200 PRINT
```

```
210 PRINT "PRESTRESSED SECTION DIMENSIONS ARE";
220 PRINT
230 GO TO 410
240 INPUT A(1),A(2),A(3),A(4),A(5),A(6),A(7),A(8),A(9),A(10)
250 IF A(1)=0 THEN 1800
260 FOR I=1 TO 10
270 IF A(I)=1 THEN 290
280 B(I)=A(I)
290 NEXT I
300 D1=B(1)
310 B1=B(2)
320 B2=B(3)
330 T1=B(4)
340 T4=B(5)
350 T2=B(6)
360 T5=B(7)
370 T3=B(8)
380 Q3=B(9)
390 Q2=B(10)
400 GO TO 950
410 INPUT A(1),A(2),A(3),A(4),A(5),A(6),A(7),A(8)
420 PRINT
430 FOR I=1 TO 8
440 IF A(I)=1 THEN 460
450 C(I)=A(I)
460 NEXT I
470 D1=C(1)
480 B1=C(2)
490 B2=C(3)
500 T1=C(4)
510 T4=C(5)
520 T2=C(6)
530 T5=C(7)
540 T3=C(8)
550 Q3=0
560 Q2=0
570 GOTO 950
580 PRINT
590 PRINT
600 IF B1>B2 THEN 630
610 G=B2
620 GO TO 640
630 G=B1
640 V=INT(G/(2*20))+1
650 V=2*V-1
660 N=0
670 GO TO 810
680 B=INT(X/(2*20))+1
690 B=2*B-1
700 D=INT(0.7*Y/20)
710 Z=((V-B)/2)+2
720 FOR P=1 TO D
730 PRINT
740 FOR Q=1 TO Z
750 PRINT " ";
760 NEXT Q
770 FOR Q=1 TO B
780 PRINT "*";
790 NEXT Q
800 NEXT P
810 N=N+1
820 IF N=1 THEN 860
830 IF N=2 THEN 890
840 IF N=3 THEN 920
850 IF N=4 THEN 120
860 X=B1
```

```
870  Y=T1
880  GO TO 680
890  X=T3
900  Y=D1-T1-T2
910  GO TO 680
920  X=B2
930  Y=T2
940  GO TO 680
950  LET D2=D1-T1-T2
960  LET T8=T4-T1
970  LET B8=(B1-T3)/2
980  LET T9=T5-T2
990  LET B9=(B2-T3)/2
1000 LET A1 =B1*T1+B2*T2+D2*T3
1010 LET A1=A1+T8*B8+T9*B9
1020 LET A1=A1+Q3*Q2
1030 LET Y1=(B1*T1†2)/(2*A1)+B2*T2*(D1-T2/2)/A1+D2*T3*(T1+C2/2)/A1
1040 LET Y1=Y1+(T8*B8*(T4-2*T8/3)/A1)+(T9*B9*(D1-T2-T9/3)/A1)
1050 LET Y1=Y1-Q3*Q2†2/(2*A1)
1060 LET Y2=D1-Y1
1070 LET Y3=Y1+Q2
1080 LET I2=B1*T1†3/12+B1*T1*(Y1-T1/2)†2
1090 LET I3=B2*T2†3/12+B2*T2*(Y2-T2/2)†2
1100 LET I4=T3*D2†3/12+D2*T3*(T1+D2/2-Y1)†2
1110 LET I4=I4+Q3*Q2†3/12
1120   LET I4=I4+Q3*Q2*((Y1+Q2/2)†2)
1130 LET I1=I2+I3+I4
1140 LET I1=I1+(B8*T8†3)/36+(B9*T9†3)/36
1150 LET I1=I1+(B8*T8*(Y1-T1-T8/3)†2)+(B9*T9*(Y2-T2-T9/3)†2)
1160 IF Q1=2 THEN 1240
1170 LET A9=A1
1180 LET W4=A9*24/1000000
1190 LET M3=W2*L1†2/8
1200 LET S5=M3*1000000*Y1/I1
1210 LET S6=M3*1000000*Y2/I1
1220 LET S7=M3*1000000*Y3/I1
1230 IF Q1=1 THEN 130
1240 LET W1=A1*24/1000000
1250 LET M1=W1*L1†2/8
1260 LET W8=W4-W1
1270 LET M2=W8*L1†2/8
1280   LET S1=M1*1000000*Y1/I1
1290 LET S2=M1*1000000*Y2/I1
1300 LET S3=(M2*1000000*Y1/I1)+S5
1310 LET S4=(M2*1000000*Y2/I1)+S6
1320 LET X1=A1*(C2+S1)
1330 LET X2=A1*(C1-S1-S3)
1340 LET Z2=A1*(S1+S3-C1)/L3
1350 LET X3=A1*(C1+S2)
1360 LET X4=A1*(S2+S4-C2)/L3
1370 LET R1=Y1*A1/I1
1380 LET R2=Y2*A1/I1
1390 IF X4>X3 THEN 1760
1400 LET E2=(X1+X4)/(R1*X4-R2*X1)
1410 LET E4=(X1+X3)/(R1*X3-R2*X1)
1420 IF C1<S1+S3 THEN 1460
1430 LET E1=(X4-X2)/(R1*X4+R2*X2)
1440 LET E3=(X3-X2)/(R1*X3+R2*X2)
1450 GO TO 1520
1460 IF Z2>X1 THEN 1760
1470 LET E1=(X4+Z2)/(R1*X4-R2*Z2)
1480 LET E3=(Z2+X3)/(R1*X3-R2*Z2)
1490 IF C1<(S1+S3) THEN 1520
1500 LET E1=I1/(A1*I1)
1510 LET E3=I1/(A1*I1)
1520 PRINT
```

```
1530 PRINT "SOLUTION POSSIBLE"
1540 PRINT
1550 PRINT "IF SOLUTION REQUIRED TYPE 1 ELSE TYPE 0";
1560 INPUT Z9
1570 PRINT
1580 IF Z9=0 THEN 120
1590 PRINT "LIMITING ECCENTRICITIES"
1600 PRINT
1610 PRINT "E1="E1"MM","E2="E2"MM"
1620 PRINT "                            Y2="Y2"MM"
1630 PRINT "E3="E3"MM","E4="E4"MM"
1640 PRINT
1650 PRINT "R2="R2
1660 PRINT
1670 PRINT "X4="X4
1680 PRINT
1690 LET M=(1.4*W4+1.6*W2)*L1↑2/8
1700 PRINT "REQUIRED ULTIMATE MOMENT = "M "KNM"
1710 PRINT
1720 PRINT "IF SECTION PROFILE IS REQUIRED THEN TYPE 1 ELSE 0";
1730 INPUT F
1740 IF F=1 THEN 580
1750 GO TO 120
1760 PRINT
1770 PRINT "INCREASE SECTION DIMENSIONS"
1780 PRINT
1790 GO TO 120
1800 END
```

3.3.3.1 Block (A): Statements 100–140

The statements in this block call for the input of the basic design parameters. These are:

1. The span (m);
2. The characteristic imposed load (kN/m);
3. The allowable concrete compressive stress (N/mm²);
4. The allowable concrete tensile stress (N/mm²);
5. The prestress loss factor.

The values input for items 1 and 2 depend upon the geometry and loading of the structure in which the member is a component.

The fact that items 3 and 4 are requested by the computer implies that the designer knows what proportion of the characteristic concrete strength may be used under service load conditions. Whilst it is reasonable to expect that he will be familiar with such requirements this raises the general issue of how responsibility should be apportioned between the programmer and the designer. There are two extremes. One is to direct the computer to ask for the end result— in this case an allowable stress. The other is to request more general information on which the specific end result is based—again in this case, the characteristic concrete strength—and to arrange for the computer to reach its own conclusion on the basis of programmed CP 110 recommendations. It is clear that a program becomes increasingly more complex when the responsibility for making executive decisions is taken away from the designer.

In the early stages of a design item 5 can only be an estimate based upon past experience of similar problems. It is a quantity which must be verified later in the design process when the losses can be assessed more accurately.

3.3.3.2 Blocks (B), (C) and (F): Statements 150–570

These blocks are concerned with the input of the sets of guessed dimensions which define the shapes of the composite and the prestressed sections. The value taken by a switch (i.e. Q1 equal to 1 or 2) determines whether Block (B) or Block (F) is entered, and consequently which of the two sections is to be dealt with.

An I-section of the shape usually associated with prestressed concrete (see Fig. 3.4a) can be defined by a string of eight dimensions. In the program these are called D1, B1, B2, T1, T4, T2, T5 and T3 respectively. They are always entered into the computer in this order. A composite section having in situ concrete standing proud of the top surface of its prestressed beam component (see Fig. 3.4b) requires two more dimensions, Q3 and Q2, for its complete definition. Other shapes (see Figs. 3.4c and d) may be defined by introducing specific zero values into the string. In the four typical sections defined below the variables D1, B1, B2, etc. represent non-zero values.

I-section	D1	B1	B2	T1	T4	T2	T5	T3	0	0
Composite I-section	D1	B1	B2	T1	T4	T2	T5	T3	Q3	Q2
Inverted Tee-section	D1	0	B2	0	0	T2	T5	T3	0	0
Composite inverted Tee-section (fully encased)	D1	0	0	0	0	0	0	T3	Q3	Q2

In the last of these examples the 'fully encased prestressed beam' definition implies a rectangular composite section with an overall depth of $(D1 + Q2)$ and a width equal to T3 (i.e. $Q3 = T3$).

The design procedure is one which normally requires a number of trials to be made. This involves the repeated typing of a string of eight or ten dimensions—a constant source of error. To eliminate this the input of section dimensions has been programmed in such a way that for all trials other than the first one only those dimensions which are to be modified need to be typed out fully. Each of the remaining dimensions may be represented by the numeral 1. The merit of this approach is that frequently only one or two dimensions are modified at each trial, consequently the volume of input is reduced.

The ten composite section dimensions are read into the elements of an array called A( ) at line 240. Line 250 comprises Block (C) which allows an exit from the program if the dimension D1 is input as zero. The contents of the array A( ) elements are examined at lines 260 to 290. If the content of an element is greater than 1, then that value is copied into the corresponding element of array B( ). Otherwise the content of the B( ) array element remains unchanged. In this way the contents of B( ) are updated to represent the current state of the

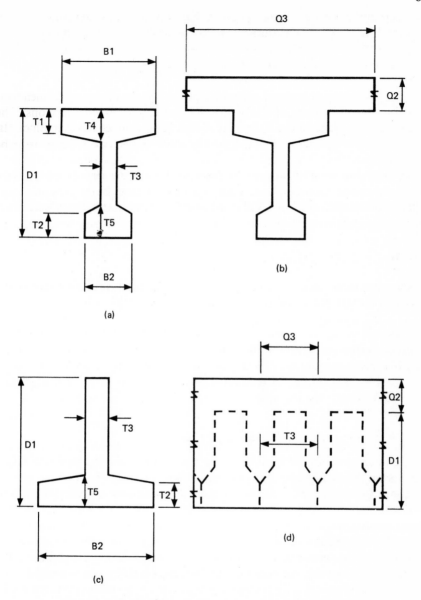

Figure 3.4 Section dimensions is terms of program PB4 variables

guessed section dimensions. Following this the variables D1, B1, B2, etc. take the values of B(1), B(2), B(3), etc. at lines 300 to 390.

In a similar operation carried out at Block (F) (lines 410 to 570) D1, B1, B2, etc. take the values of the prestressed section dimensions. It will be noted that there are now only eight dimensions to be read into A( ) and that Q3 and Q2 are given zero values at lines 550 and 560.

Using this sub-program the standard design of a prestressed beam may be accomplished by reading in the same set of dimensions at the composite section stage (with the exception of an additional pair of zeros to represent the now non-existent dimensions Q3 and Q2) as for the prestressed beam itself.

3.3.3.3 Block (D): Statements 950–1160

The section properties calculated at this stage are governed by the current value of Q1 which determines whether the composite section or the prestressed section is processed. The properties of interest are the area of the cross-section (A1), the distances of the upper and lower surfaces of the section to the centre of area (Y1 and Y2 respectively) and the second moment of area (I1). The calculation of each section property is carried out in a number of stages for ease of typing and checking.

3.3.3.4 Blocks (E) and (I): Statements 1170–1310

The purpose of these two blocks is to calculate the bending stresses in the outer fibres of the prestressed beam component according to whether the section is currently a composite one (i.e. Q1 = 1 at Block (E)—Statements 1170 to 1230) or a prestressed beam acting on its own (i.e. Q1 = 2 at Block (I)—Statements 1240 to 1310). In Block (E) S5 and S6 (i.e. the stresses c_{st} and c_{sb} which were defined in Section 3.2) are calculated at lines 1200 to 1210. In Block (I) S1, S2, S3 and S4 (i.e. the stresses c_{dt}, c_{db}, $(c_{dt} + c_{st})$ and $(c_{db} + c_{sb})$ which were also defined in Section 3.2) are calculated at lines 1280 to 1310.

3.3.3.5 Blocks (J) and (K): Statements 1320–1510

The function of Block (J) is to check the position of the line representing inequation (3.1) relative to that of inequation (3.2a) or (3.2b), and the relative positions of lines (3.3) and (3.4), in order to determine whether the current trial sections offer a possible solution. If these pairs of lines are found to be in their correct order then the boundary values to the eccentricities of the prestressing force are calculated.

Recasting the inequations (3.1) to (3.4) of Section 3.2 by substituting for other than P and e the names of the variables used in the program gives:

(3.1) becomes $\dfrac{1}{P} \geq \dfrac{(e*R1 - 1)}{X1}$

(3.2a) becomes $\dfrac{1}{P} \geq \dfrac{(1 - e*R1)}{X2}$

(3.2b) becomes $\dfrac{1}{P} \leq \dfrac{(e*R1 - 1)}{Z2}$

(3.3) becomes
$$\frac{1}{P} \geqq \frac{(e*R2 + 1)}{X3}$$

and (3.4) becomes
$$\frac{1}{P} \leqq \frac{(e*R2 + 1)}{X4}$$

where $R1 = y_1/r^2$, $R2 = y_2/r^2$, $X1 = A(c_t + c_{dt})$, $X2 = A(c - c_{dt} - c_{ct} - c_{st})$, $Z2 = A(c_{dt} + c_{ct} + c_{st} - c)$, $X3 = A(c + c_{db})$ and $X4 = A(c_{db} + c_{cb} + c_{sb} - c_t)$.

The contents of the variables R1, R2, X1, X2, Z2, X3 and X4 are calculated at lines 1320 to 1380.

To check whether lines (3.3) and (3.4) occur in their correct order the values of their intercepts with the $1/P$ axis are calculated and compared. For $e = 0$ the intercepts of lines (3.3) and (3.4) are $1/X3$ and $1/X4$ respectively. These lines are correctly related if $1/X4$ is $\geqq 1/X3$, or if X4 is $<$ X3. This relationship is checked at line 1390. If X4 is found to be greater than X3 then a 'section fail' message is printed out at line 1770. Otherwise, since the section has at least proved itself on this account to be theoretically large enough to furnish a solution, the limiting eccentricities E2 and E4 are calculated.

Referring to Fig. 3.2, the lines (3.1) and (3.4) intersect at an eccentricity E2 which represents the maximum allowable eccentricity for the prestressing force in that section. Equating the RHS's of equations (3.1) and (3.4) gives:

$$\frac{(E2*R1 - 1)}{X1} = \frac{(E2*R2 + 1)}{X4}$$

or that $E2 = (X1 + X4)/(R1*X4 - R2*X1)$.

This calculation is carried out at line 1400. The intersection of lines (3.1) and (3.3) gives E4 at program line 1410.

In a similar way to that already discussed the relationship between lines (3.3) and (3.2a) or (3.3) and (3.2b) is examined. If this still shows the section to be large enough to yield a solution then the remaining limiting eccentricities E1 and E3 are calculated either at lines 1430 and 1440 or lines 1470 and 1480 depending upon whether c is greater or less than $(c_{dt} + c_{ct} + c_{st})$.

The four exits from Block (J) comprise Block (K) in the flow diagram. If a section inadequacy is revealed at lines 1390 or 1460 then the program jumps to line 1760 where the designer is advised to increase the section dimensions. Jumps from lines 1450 and 1490 to line 1520 indicate that a solution is possible.

3.3.3.6 Blocks (L), (G) and (M): Statements 1520–1800

Block (L) is entered at line 1520. At an early stage in the design procedure it is often sufficient for the designer to know that a solution is possible without his needing the detailed results. The statements at lines 1530 to 1580 make this possible.

A normal output of results consists of a summary of the limiting eccentricities E1 to E4 (from which the designer will choose an actual eccentricity for the prestressing force), the values of Y2, R2 and X4 (further quantities needed when the prestressing force is eventually calculated) and the ultimate moment of resistance required of the section (for later reference when Sub-program PB5 is run).

In addition to these results, at Block (G)—see lines 1720 to 1750—the designer may opt for a pictorial representation of the section. This procedure is described in Section 3.3.3.7.

Block (M) comprises the statements at lines 1760 to 1790. If this block is entered then the designer is informed that the section is too small and a jump back to line 120 gives him the opportunity to revise the section dimensions.

3.3.3.7 Block (H): Statements 580–940

The object of this part of the program is to give a pictorial representation of the prestressed beam component of the composite section. This is achieved by printing out blocks of asterisks to represent the relative sizes and positions of the flanges and web of the section. Each block of asterisks is defined by the variables B and D, the number of asterisks in a row and the number of rows in a block respectively. The expressions for calculating B and D (see lines 680 to 700) were arrived at empirically to give a figure of approximately $\frac{1}{8}$th scale. Flanges are reproduced as rectangular blocks; at this scale the increment of depth represented by one newline is too coarse to reproduce the inner sloping surface. The position of the first asterisk in a row is controlled by the number of spaces, Z (see line 710), which precedes it. Thus it is possible to locate the rows symmetrically about the section centreline and to build up a picture by suitably varying the parameters Z, B and D.

A variable called N, which is set to zero at line 660, takes the value of $(N + 1)$ each time the procedure passes line 810. The function of N is to select the three basic parts of the section in their correct order. For $N = 1$, say, the dimensions of the first block of asterisks are based upon an actual width of $X = B1$ and an actual depth of $Y = T1$—i.e. the top flange—at lines 860 and 870. A jump to line 680 converts these full scale dimensions into their $\frac{1}{8}$th. scale equivalents in terms of a number of asterisks (see B at lines 680 and 690) or newlines (see D at line 700). The content of Z, which is calculated at line 710, ensures that the block of asterisks will be correctly located relative to the centreline of the block of greatest width—i.e. V, calculated at lines 600 to 650. The printout procedure at lines 720 to 800 defines a block of asterisks in terms of P (lines 720, 730 and 800) which controls the depth, Q (lines 740 to 760) which governs the number of spaces to appear at the beginning of a newline and Q (lines 770 to 790) which controls the number of asterisks per line. The whole procedure is carried out for N equal successively to 1, 2 and 3. When N is equal to 4 the statement at line 850 provides an exit from Block (H).

By using techniques which were similar to those described in this section a procedure was developed to show the relative spatial relationships between the limiting eccentricities and $1/P$. Since its usefulness proved to be limited this facility was deleted from the present version of PB4. Even so it is worthwhile remembering that the value of some results might be enhanced by being presented graphically.

3.3.4 Program Specification—PB2

The purpose of this sub-program is to determine the number and type of prestressing tendons which are needed for a section proportioned according to the service load bending requirements of Sub-program PB4. PB2 gives a solution based upon a prestressing system which uses tendons employing groups of stranded wires; to complete the library a similar sub-program would be needed for each favoured prestressing system.

3.3.5 Description of Sub-program PB2

A list of the variables and arrays used is given below, followed by a program listing.

A( ) Holds a description of the prestressing system and possible solutions
E1 Lower limiting eccentricity
E9 Estimated eccentricity
I Counter
J Counter

P1 1st calculated value of prestressing force
R2 Output from PB4
T Web thickness
W1 Worth of one tendon before losses
X4 Output from PB4
Y2 Output from PB4

```
100 DIM A(11,11)
110 FOR I=1 TO 11
120 FOR J=1 TO 5
130 READ A(I,J)
140 NEXT J
150 NEXT I
160 DATA 7,4,12.5,94.2,1523
170 DATA 7,4,15.2,138.7,1423
180 DATA 19,4,18,210,1532
190 DATA 7,7,12.5,94.2,1523
200 DATA 7,7,15.2,138.7,1423
210 DATA 19,7,18,210,1532
220 DATA 7,12,12.5,94.2,1523
230 DATA 7,12,15.2,138.7,1423
240 DATA 7,15,15.2,138.7,1423
250 DATA 7,19,15.2,138.7,1423
260 DATA 19,19,18,210,1532
270 FOR I=1 TO 11
280 READ A(I,11)
290 NEXI I
300 DATA 45,55,70,60,70,80,70,80,90,110,110
310 PRINT "WEB THICKNESS OF SECTION (MM)=";
320 INPUT T
330 PRINT
340 IF T<140 THEN 910
```

```
350 FOR I=1 TO 11
360 IF T>=3*A(I,11) THEN 380
370 LET A(I,11)=0
380 NEXT I
390 PRINT "LOWER LIMITING ECCENTRICITY E1 (MM)=";
400 INPUT E1
410 PRINT
420 PRINT "ESTIMATED ECCENTRICITY OF PRESTRESSING FORCE (MM) =";
430 INPUT E9
440 PRINT "VALUES OF Y2,R2 AND X4 TAKEN FROM"
450 PRINT "PROGRAMME PB4 OUTPUT ARE";
460 INPUT Y2,R2,X4
470 PRINT
480 PRINT
490 LET P1=X4/((1+E9*R2)*1000)
500 PRINT
510 PRINT "PRESTRESSING FORCE FOR ESTIMATED ECCENTRICITY="F1"KN"
520 PRINT
530 PRINT
540 FOR I=1 TO 11
550 IF A(I,11)=0 THEN 650
560 LET W1=A(I,2)*A(I,4)*A(I,5)*0.7/1000
570 LET A(I,6)=INT(P1/W1)+1
580 LET A(I,8)=A(I,6)*W1
590 LET A(I,7)=(X4/(A(I,8)*1000)-1)/R2
600 IF A(I,7)>=E1 THEN 630
610 LET A(I,11)=0
620 GOTO 650
630 LET A(I,9)=INT((A(I,8)-P1)*100/P1)+1
640 LET A(I,10)=Y2-A(I,7)
650 NEXT I
660 PRINT "OUT OF THE 11 STRANDED CABLES CONSIDERED"
670 PRINT "THE FOLLOWING CABLES CAN BE ACCOMMODATED IN THE WEB ********"
680 PRINT
690 PRINT
700 PRINT "CABLE    NO. OF    NO. OF     NOMINAL"
710 PRINT "REF.     WIRES/    STRANDS/   STRAND"
720 PRINT "NO.      STRAND    CABLE      DIA.(MM)"
730 PRINT
740 FOR I=1 TO 11
750 IF A(I,11)=0 THEN 770
760 PRINT " "I"      "A(I,1)"      "A(I,2)"          "A(I,3)
770 NEXT I
780 PRINT
790 PRINT
800 PRINT "POSSIBLE CABLE GROUPS ************"
810 PRINT
820 PRINT"CABLE ACTUAL              NO. OF STEEL    DIST. FROM ETM."
830 PRINT"REF.  PRES.    %AGE   CABLES SERVICE FLANGE TO C.G. OF"
840 PRINT"NO.   FORCE(KN) EXCESS REQD,  STRESS  CABLE SYSTEM (MM)"
850 PRINT
860 FOR I=1 TO 11
870 IF A(I,11)=0 THEN 890
880 PRINT" "I" "A(I,8)"   "A(I,6)"   "0.7*A(I,5)"     "A(I,10)
890 NEXT I
900 GOTO 930
910 PRINT "MINIMUM ALLOWABLE WEB THICKNESS FOR THIS "
920 PRINT "CABLE SYSTEM=140 MM"
930 END
```

Whilst the form of the program is too simple to merit a flow diagram, for the purposes of discussion it is convenient to divide the statements into four blocks; these are discussed in the following sub-sections.

3.3.5.1 Block (A): Statements 100–300

The initial role of array A( ) is to store a numerical description of the prestressing system. This information is conveyed to the computer via the READ and DATA statements at lines 110 to 300. Each row in the array relates to a specific stranded tendon for which information is entered in the following order:

J = 1, the number of wires per strand;
J = 2, the number of strands per tendon;
J = 3, the nominal strand diameter;
J = 4, the cross-sectional area of one strand;
J = 5, the design strength of one strand;
J = 11, the required duct diameter for this tendon.

3.3.5.2 Block (B): Statements 310–380

When setting an upper limit to the choice of tendon size the web thickness of the section is the most restrictive dimension. In Block (B) it is considered that the web thickness must at least be equal to (3∗duct diameter) before that tendon may even be considered on other grounds. This also implies that for a given prestressing system there will be a minimum web thickness which is capable of housing the smallest tendon. The statement at line 340 checks this and provides a jump to the end of the program should the web prove to be too thin.

The procedure at lines 350 to 380 identifies each of the tendons which may possibly be used from the list of those available. This is achieved by comparing the web thickness with each duct diameter in turn. If a duct is too large to be accommodated the fact is recorded in array A( ) by setting the content of A(I,11) to zero. In subsequent calculations, whenever it is noted that A(I,11) = 0 then that tendon is ignored.

3.3.5.3 Block (C): Statements 390–530

Since the line representing inequation (3.4) (see Fig. 3.2) gives a range of eccentricities for which the prestressing force is at a minimum, the estimated eccentricity (E9) which is input at line 430 should be chosen from the range between E1 and E2. At the best the value of E9 can only be an educated guess because at the time the estimate is made the proportion of the beam depth taken up by the tendons is unknown. Some guidance in estimating E9 can be obtained by comparing the two dimensions E2 and Y2 given in the PB4 output.

Restating inequation (3.4), Section 3.3.3.5, in terms of the Sub-program PB2 variables gives

$$\frac{1}{P} \leq \frac{(E9 \ast R2 + 1)}{X4}.$$

This leads to the statement at line 490 which gives the prestressing force (P1) in terms of the estimated eccentricity (E9).

3.3.5.4 Block (D): Statements 540–930

Since it is unlikely that P1 will translate directly into a whole number of tendons working at an allowable stress equal to (0.7∗the design strength), the area of prestressing steel provided will always be greater than that theoretically required. Of the options this situation presents, the programmed calculation translates the rounded-up number of tendons—see lines 560 and 570—into a larger prestressing force (larger because the steel is still assumed to be tensioned up to its allowable stress) placed at a smaller eccentricity than that initially estimated. This has the effect of giving the designer greater latitude in positioning the tendons at the expense of a reduction in the ultimate moment of resistance of the beam. These calculations take place at lines 540 to 650 where, according to the statement at line 550, only those tendons which are capable of being accommodated in the web are considered. For each of the possible stranded tendons in the list the rounded-up number of tendons required to develop P1 is calculated at line 560 and converted to an actual prestressing force—called A(I, 6)—at line 570. The corresponding eccentricity A(I, 7) follows from the statement at line 590. A check at line 600 compares the content of A(I, 7) with the minimum allowable eccentricity E1 and rejects the tendon as a possibility if A(I, 7) is less than E1. The percentage excess of the actual prestressing force over that initially estimated is calculated at line 630. This serves to indicate the efficiency of the tendon in terms of the area of steel used relative to that initially estimated.

The remainder of Block (D) is concerned with the output of information of interest to the designer. This includes a list of the properties of possible stranded tendons together with further detailed information which will assist him in making a choice.

3.3.6 Program Specification—PB3

Sub-program PB3 has a threefold purpose:

1. To offer guidance concerning the maximum number of tendons that may be accommodated across the width of the section;
2. To calculate the equal distance between rows of tendons for which the resultant of the system coincides with the required eccentricity;
3. To give the minimum width of end-block which will accommodate the anchorages.

This sub-program is orientated towards a particular stranded tendon prestressing system. A complete library would require a series of similar sub-programs so that the whole range of available systems could be considered.

3.3.7 Description of Sub-program PB3

A list of the variables and arrays used in it is given below, followed by a program listing.

A	Width of bottom flange	N4	Total number of tendons
B( )	Holds a description of the prestressing system	N6	Largest number of tendons/row
		N8	Maximum number of tendons across flange
C	Web thickness		
D	Strand diameter of proposed tendon	N9	Maximum number of tendons across web
I	Counter	S1	Distance of system c.g. from 1st row, for unit row spacing
I1	Counter		
J	Counter		
M1	Moment of tendons about the c.g. of the 1st row	S2	Distance from bottom of beam to c.g. 1st row
N	Number of strands/tendon	S3	Distance from bottom of beam to c.g. cable system
N1	Number of rows		
N2	Row number	S4	Calculated distance between rows
N3	Number of tendons in successive rows	Z	Switch

```
100 DIM B(11,5)
110 FOR I=1 TO 11
120 FOR J=1 TO 5
130 READ B(I,J)
140 NEXT J
150 NEXT I
160 DATA 4,12.5,45,140,150
170 DATA 4,15.2,55,165,200
180 DATA 4,18,70,190,255
190 DATA 7,12.5,60,165,200
200 DATA 7,15.2,70,190,255
210 DATA 7,18,80,230,280
220 DATA 12,12.5,70,150,265
230 DATA 12,15.2,80,200,330
240 DATA 15,15.2,90,215,355
250 DATA 19,15.2,110,230,380
260 DATA 19,18,110,285,495
270 PRINT "PROPOSED CABLE - NUMBER OF STRANDS/CABLE"
280 PRINT "AND STRAND DIAMETER (MM) ARE";
290 INPUT N,D
300 PRINT
310 FOR I1=1 TO 11
320 LET I=I1
330 IF N*D=B(I1,1)*B(I1,2) THEN 350
340 NEXT I1
350 PRINT "BOTTOM FLANGE WIDTH (MM)=";
360 INPUT A
370 PRINT "WEB THICKNESS (MM)=";
380 INPUT C
390 PRINT
400 LET N8=INT((A-B(I,3))/(2*B(I,3)))
410 LET N9=INT((C-B(I,3))/(2*B(I,3)))
420 PRINT "MAXIMUM NUMBER OF CABLES ACROSS FLANGE="N8
430 PRINT "MAXIMUM NUMBER OF CABLES ACROSS WEB="N9
440 PRINT
```

```
450 PRINT "DECIDE UPON CABLE ARRANGEMENT ***********"
460 PRINT
470 PRINT
480 PRINT "HOW MANY ROWS";
490 INPUT N1
500 PRINT "DISTANCES (MM) FROM BOTTOM OF BEAM"
510 PRINT "TO C.G. OF FIRST ROW AND C.G. OF CABLE GROUP ARE"
520 INPUT S2,S3
530 LET N4=0
540 LET M1=0
550 LET N2=0
560 LET N6=0
570 PRINT "NUMBER IN ROW"(N2+1):
580 INPUT N3
590 IF N3<=N6 THEN 610
600 LET N6=N3
610 LET N2=N2+1
620 LET N4=N4+N3
630 LET M1=M1+N3*(N2-1)
640 IF N2<N1 THEN 570
650 LET S1=M1/N4
660 LET S4=(S3-S2)/S1
670 PRINT
680 PRINT "DISTANCE TO CENTRE OF FIRST ROW IS"S2"MM"
690 PRINT
700 PRINT "DISTANCE BETWEEN CENTRES OF ROWS IS"S4"MM"
710 PRINT
720 PRINT "MINIMUM END BLOCK WIDTH TO ACCOMMODATE"
730 PRINT "ANCHORAGES="(N6-1)*B(I,5)+2*B(I,4)"MM"
740 PRINT
750 PRINT "IF ALL ARRANGEMENTS HAVE BEEN CONSIDERED"
760 PRINT "THEN TYPE 1 ELSE 0";
770 INPUT Z
780 PRINT
790 IF Z=0 THEN 270
800 END
```

For the purposes of discussion the sub-program is divided into three blocks of statements; these are discussed in the following sub-sections.

3.3.7.1 Block (A): Statements 100–260

Array B( ) holds the numerical information which describes the properties of the prestressing system relevant to the calculations in this sub-program. Each row in the array relates to a particular stranded tendon for which the information is held as follows:

$J = 1$, the number of strands/tendon;
$J = 2$, the nominal strand diameter;
$J = 3$, the duct diameter required for this tendon;
$J = 4$, the minimum allowable dimension from the centre of the anchorage to the edge of the end-block;
$J = 5$, the minimum allowable dimension between the centres of anchorages.

3.3.7.2 Block (B): Statements 270–440

A tendon type will have been selected by the designer from the information output by Sub-program PB2. This choice is conveyed to the computer by assigning appropriate values to the variables N and D at lines 270 to 290. A search is now made through the rows of array B( ) to locate the position of the chosen tendon—see lines 310 to 340. At each row in the array the products $N*D$ and $B(I1,1)*B(I1,2)$ are compared. Because $N*D$ uniquely defines one tendon in the system, the row at which the two products coincide is the required location; the location (I1) of this row is permanently recorded in the variable called I at line 320.

To aid the designer in deciding upon the pattern of tendons in the section the calculations at lines 350 to 430 determine the maximum number of tendons that it is possible to accommodate within the widths of the web and bottom flange. These assessments are based upon edge covers and clear spaces between ducts both equal to the duct diameter $B(I, 3)$.

3.3.7.3 Block (C): Statements 450–800

With the information output from Block (B) the designer is able to decide upon a possible arrangement of tendons. The total number of tendons required and the position of the c.g. of the system were output from PB2. This information, together with the designer's estimate of the distance to the c.g. of the first row, the number of rows and the number of tendons in each row comprise the input on which the row spacing calculation is based. The variables N4, M1, N2 and N6 are set to zero at lines 530 to 560, a necessary precaution if other arrangements are likely to be considered during the same program run. Considering each row in turn the procedure at lines 570 to 640 executes the following operations: the number of tendons in the row currently being considered is input at line 580; the statements at lines 590 and 600 record the largest number of cables in a single row in the variable N6, information which is used later to determine the minimum end-block width; at line 610 the variable N2 records the number of rows so far considered and at line 620 a running total is kept in N4 of the number of tendons; following this, M1 records the sum of the 1st moments of each row of tendons about the first row (this assumes unit distance between rows) at line 630; the statement at line 640 returns the calculation to a further row input until all the rows have been considered. Using the results of this procedure the statements at lines 650 and 660 give a distance (S4) between rows such that the resultant of the system coincides with the required eccentricity.

Information which defines the position of each row relative to the outer edge of the bottom flange is output at lines 680 to 700. The final item of output information concerns the minimum width of end-block which is necessary to house the anchorages; this is calculated and output at line 730.

Further tendon arrangements may be considered by typing in the appropriate response at line 750.

3.3.8 Program Specification—PB5

Working with data which defines the dimensions and material properties of a prestressed concrete beam that has already been proportioned to meet service load bending requirements, this sub-program determines its moment of resistance at the ultimate limit state of bending.

3.3.9 The PB5 Flow Diagram

The flow diagram (see Fig. 3.5) shows the six areas into which this sub-program has been divided for the purposes of discussion.

Following the input of data at Blocks (A) and (B) an iterative procedure defined by the loop

$$\overset{\longleftarrow}{C \rightarrow D \rightarrow E}$$

establishes equilibrium between the resultant tensile and compressive forces which act at the section. On entering Block (C) for the first time it is arbitrarily assumed that the tendon steel has fully yielded. This permits a (probably approximate) value to be assigned to the ultimate force developed by the tendons and provides a basis on which to estimate the depth of neutral axis necessary for internal equilibrium. But the steel strains implied by this neutral axis depth are not necessarily compatible with the steel stress which was assumed. Block (D) is therefore concerned with calculating the total steel strain at the level of each row of tendons, relating this to the steel stress/strain properties and hence re-estimating the resultant tensile force. The estimated values of the tensile force at the beginning and end of an iteration are compared at Block (E); if the difference between them is unacceptably large then the loop is re-cycled. The second and subsequent iterations take an identical form to that already described with the exception that the value of tensile force used to determine a new neutral axis depth is taken as the average of the two values which were compared in Block (E) at the conclusion to the previous iteration. This is an effective way of damping down any tendency for the solution to oscillate. An exit is made from Block (E) once a stable value of the tensile force, and hence the neutral axis depth, has been established. This is followed by the calculation and output at Block (F) of information concerning the ultimate moment of resistance of the section.

3.3.10 Description of Sub-program PB5

A list of variables and arrays used in this sub-program is given below.

A( )	Contains the number of cables in each row	I	Counter
A1	Area of concrete in compression	I1	Row identifier
		J	Counter
A2	Rectangular portion of flange	L1	Lever arm
		M9	Required ultimate moment
		M8	Difference between actual and required ultimate moment

A3	Total area of flange
A4	Step in calculation of D9
B( )	Contains distances between rows
B1	Section dimension
B2	Section dimension
B9	Step in calculation of D9
C( )	Contains the steel strain in each row
C1	Characteristic concrete strength
D( )	Contains the steel stress in each row
D1	Section dimension
D7	Distance between rows
D8	Distance to row O2
D9	Distance to c.g. of A1
E1	Steel strain due to bending
E2	Steel strain due to prestress after losses
E3	Total steel strain
F1, F2, F3, F4	See Fig. 3.6
F9	Percentage of steel design strength
G( )	Contains a description of the prestressing system

N1	Number of tendons in row O2
N9	Number of tendons
O1	Number of rows
P1	Tendon force in row O2
P2	Tensile force
Q1	Distance to c.g. of tensile force
Q7	Distance from bottom of beam to 1st row
Q8	Distance between rows
S1	Strand diameter
S9	Steel stress due to total strain
T1, T2, T3, T4	Section dimensions
T9	Tensile force
U1	Ultimate moment of resistance of section
X1	Distance to NA from top of flange
X2	Intermediate step to X1
Y8	Loss factor
Y9	Number of strands/tendon
Z1, Z2	Intermediate steps to X1
O2	Counter

Program PB5 is listed below and should be read in conjunction with the flow diagram shown in Fig. 3.5.

```
100 DIM G(11,5)
110 FOR I=1 TO 11
120 FOR J=1 TO 5
130 READ G(I,J)
140 NEXT J
150 NEXT I
160 DATA 4,12.5,94.2,1523,200
170 DATA 4,15.2,138.7,1423,200
180 DATA 4,18,210,1532,175
190 DATA 7,12.5,94.2,1523,200
200 DATA 7,15.2,138.7,1423,200
210 DATA 7,18,210,1532,175
220 DATA 12,12.5,94.2,1523,200
230 DATA 12,15.2,138.7,1423,200
240 DATA 15,15.2,138.7,1423,200
```

```
250 DATA 19,15.2,138.7,1423,200
260 DATA 19,18,210,1532,175
270 PRINT "REQUIRED MULT (KNM)=";
280 INPUT M9
290 PRINT "YOUR VALUES OF OVERALL BEAM DEPTH,"
300 PRINT "UPPER FLANGE WIDTH,"
310 PRINT "OUTER AND INNER FLANGE THICKNESSES"
320 PRINT "AND WEB THICKNESS (MM) ARE"
330 INPUT D1,B1,T1,T2,T3
340 PRINT "CHARACTERISTIC CONCRETE STRENGTH (N/MM+2)"
350 PRINT "AND LOSS FACTOR ARE";
360 INPUT C1,Y8
370 PRINT "NUMBER OF CABLES=";
380 INPUT N9
390 PRINT "CABLE TYPE - NUMBER OF STRANDS/CABLE"
400 PRINT "AND STRAND DIAMETER (MM) ARE";
410 INPUT Y9,S1
420 FOR I=1 TO 11
430 I1=I
440 IF Y9*S1=G(I,1)*G(I,2) THEN 460
450 NEXT I
460  PRINT "HOW MANY ROWS";
470 INPUT O1
480 FOR I=1 TO O1
490 PRINT "NUMBER OF CABLES IN ROW"I;
500 INPUT A(I)
510 NEXT I
520 PRINT "DISTANCE (MM) FROM BOTTOM OF BEAM"
530 PRINT "TO C.G. OF FIRST ROW=";
540 INPUT O7
550 B(1)=Q7
560 PRINT "DISTANCE (MM) BETWEEN ROWS=";
570 INPUT Q8
580 FOR I=2 TO O1
590 B(I)=Q8
600 NEXT I
610 B2=(B1-T3)/2
620 T4=T2-T1
630 A2=B1*T1
640 A3=A2+B2*T4+T3*T4
650 T9=N9*G(I1,1)*G(I1,3)*G(I1,4)
660 GOTO 680
670 T9=(P2+T9)/2
680 A1=T9/(0.4*C1)
690 IF A1<=A2 THEN 720
700 IF A1<=A3 THEN 750
710 IF A1>A3 THEN 820
720 X1=A1/B1
730 D9=X1/2
740 GOTO 880
750 Z1=B1*T4/B2
760 Z2=T4*(A1-A2)/B2
770 X2=(Z1-SQR(Z1+2-4*Z2))/2
780 X1=X2+T1
790 B9=T3+2*(T4-X2)*B2/T4
800 D9=((B1*T1+2)/2+B9*X2*(T1+X2/2)+X2*(B1-B9)*(T1+X2/3)/2)/A1
810 GOTO 880
820 A4=A1-A3
830 X2=A4/T3
840 X1=X2+T2
850 D9=(B1*T1+2)/2+T4*B2*(T1+T4/3)
860 D9=D9+T4*T3*(T1+T4/2)+X2*T3*(X2/2+T2)
870 D9=D9/A1
880 E2=0.7*G(I1,4)*Y8/(G(I1,5)*1000)
890 D8=0
900 P2=0
```

```
910 Q1=0
920 O2=0
930 O2=O2+1
940 N1=A(O2)
950 D7=B(O2)
960 D8=D8+D7
970 F1=0.8*G(I1,4)/(G(I1,5)*1000)
980 F2=0.005+G(I1,4)/(G(I1,5)*1000)
990 F3=F2-F1
1000 F4=0.2*G(I1,4)
1010 E1=0.0035*(D1-X1-D8)/X1
1020 E3=E2+E1
1030 IF E3>=F2 THEN 1090
1040 IF E3 >=F1 THEN 1070
1050 S9=E3*G(I1,5)*1000
1060 GOTO 1100
1070 S9=0.8*G(I1,4)+(E3-F1)*F4/F3
1080 GOTO 1100
1090 S9=G(I1,4)
1100 P1=S9*G(I1,1)*G(I1,3)*N1
1110 P2=P2+P1
1120 Q1=Q1+P1*D8
1130 C(O2)=E3
1140 D(O2)=S9
1150 IF O2<O1 THEN 930
1160 IF ABS(T9-P2)>0.001*P2 THEN 670
1170 Q1=Q1/P2
1180 L1=D1-D9-Q1
1190 U1=L1*P2/1000000
1200 M8=U1-M9
1210 IF M8>0 THEN 1250
1220 PRINT "BEAM M. ULT. LESS THAN REQUIRED ************"
1230 PRINT "INCREASE SECTION DIMENSIONS ****************"
1240 GOTO 1390
1250 PRINT
1260 PRINT "ACTUAL MULT GREATER THAN REQUIRED"
1270 PRINT
1280 PRINT "ULTIMATE MOMENT OF RESISTANCE="U1"KNM"
1290 PRINT
1300 PRINT "DEPTH OF NEUTRAL AXIS="X1"MM"
1310 PRINT
1320 PRINT "STEEL STRESS AT ULTIMATE LOAD EXPRESSED AS A"
1330 PRINT "PERCENTAGE OF THE DESIGN STRENGTH"
1340 PRINT
1350 FOR I=1 TO O1
1360 F9=INT(D(I)*100/G(I1,4))
1370 PRINT "AT ROW"I"    "F9"%"
1380 NEXT I
1390 END
```

3.3.10.1 Block (A): Statements 100–260

Each row in array G( ) holds numerical information which relates to a specific stranded cable and its steel properties. This information is held by the row in the following order:

$J = 1$, the number of strands/tendon;
$J = 2$, the nominal strand diameter;
$J = 3$, the cross-sectional area of one strand;
$J = 4$, the steel design strength;
$J = 5$, the Young's modulus.

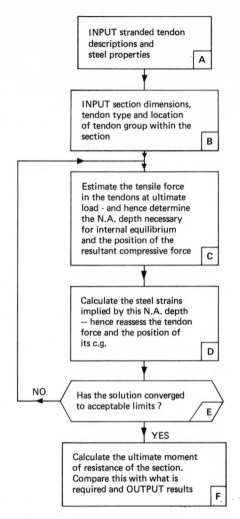

Figure 3.5 Flow diagram for program PB5 (analysis of prestressed concrete section at the ultimate limit state of bending)

3.3.10.2 Block (B): Statements 270–600

The ultimate moment of resistance required of the section—a function of the beam span and its loading—was given in the Sub-program PB4 output. This value is input to Sub-program PB5 at line 280 in order that, at a later stage, a comparison may be made with the actual moment of resistance.

Since it is not initially known where the neutral axis will lie it is necessary to define the change in shape of the section so that calculations appropriate to the actual shape of the compression area may be made regardless of the neutral axis position. The section dimensions, input at lines 290 to 330, are followed by the input of the characteristic concrete strength and prestress loss factor at lines 340 to 360.

The remainder of this block of statements is concerned with the input of information which describes the type and number of tendons and the levels they occupy in the section. A procedure at lines 390 to 450 (cf. Section 3.3.7.2) locates the row in array G( ) at which the relevant tendon properties are held. This row number is stored in the variable I1 for later reference. The number of tendons in each row is input to array A( ) at lines 460 to 510 and the spacing between the rows is input to array B( ) at lines 520 to 600.

3.3.10.3 Block (C): Statements 610–870

The purpose of this part of the sub-program is to find the neutral axis position. This is done by equating the resultant compressive force (which is itself in terms of the depth of the compressive stress block, i.e. the neutral axis depth) to successively better estimates of the tensile force.

An estimate of the tensile force (T9) is made either at line 650 or at line 670 depending upon whether the calculation is in its first iteration or a later one. The initial estimate assumes that the tendons have fully yielded. If, subsequently, this proves not to be the case then an improved estimate of T9 is made at line 670.

A1, the area of concrete which is required to be in compression for there to be equilibrium with the tensile force T9, is calculated at line 680. This assumes that at the ultimate limit state of bending the compressive stress is uniformly distributed and equal to $0.4f_{cu}$. At lines 690 to 710 A1 is compared with A2 and A3, the area of the upper rectangular portion of the flange and the total flange area respectively. Thus a conclusion is formed regarding the general region within which the neutral axis lies. The calculation is then routed to one of the three program areas headed by the lines 720, 750 or 820 where the appropriate values of X1 (the neutral axis depth) and D9 (the distance from the outer edge of the top flange to the resultant compressive force) are found.

3.3.10.4 Block (D): Statements 880–1150

One component of the total steel strain is E2, the strain that results when the tendons are stretched to induce a prestressing force. This is calculated at line 880 where the recorded value takes account of the estimated losses in prestress. E2 is a blanket value which is assigned to all tendons regardless of their position in the section.

Within the loop defined by lines 930 to 1150 each row of tendons is considered in turn by advancing the value of O2 from 1 to O1, the total number of rows. At lines 940 to 960 the variables N1 and D8 take, respectively, the values of the number of tendons in row O2 and the distance of that row from the bottom edge of the lower flange. The steel stress/strain diagram shown in Fig. 3.6 defines

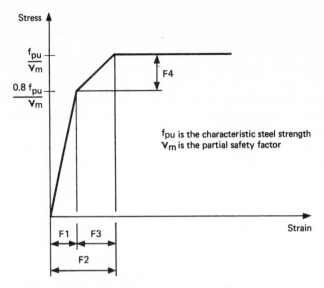

Figure 3.6 Definition of steel stress/strain diagram in terms
of program PB5 variables (see lines 970 to 1000)

significant values which are used by the sub-program to calculate stresses
from known strains. The variables F1, F2, F3 and F4 take the values assigned
to them at lines 970 to 1000.

The steel strain due to bending is defined by the content of the variable E1
at line 1010. This strain is calculated on the assumption that the maximum
concrete compressive strain is 0.0035 and that the strain distribution is linear
over the depth of the section. The total steel strain at the level of row O2 is
(E1 + E2); this is assigned to the variable E3 at line 1020.

In order to bracket the range of strain on the steel stress/strain diagram within
which E3 occurs, the total strain is compared in turn with F2 and F1 at lines
1030 and 1040. Following this the appropriate steel stress (S9) is calculated at
one of the lines 1050, 1070 or 1090.

The tendon force in row O2 is calculated and assigned to the variable P1
at line 1100. A running total of the tensile force in each row is recorded in the
variable P2 at line 1110. At line 1120 Q1 is assigned the sum of the moments of
the tensile force in each row about the lower edge of the bottom flange. The
individual strain and stress at each row of tendons is stored in arrays C( )
and D( ) at lines 1130 and 1140.

3.3.10.5 Block (E): Statement 1160

Block (E) comprises a single statement. Its purpose is to compare the values
of the tensile force (T9), which was estimated at the beginning of the current
iteration, with (P2), a value derived from considerations of strain compatability.

When these are within 0.1% of each other an exit to Block (F) follows; otherwise the loop is recycled.

3.3.10.6 Block (F): Statements 1170–1390

Since at this stage the calculated value of the tensile force (P2) has reached a stable value, equilibrium between the resultant forces which act at the section may be assumed.

The statement at line 1170 calculates Q1, the distance of the c.g. of the tensile force from the outer edge of the bottom flange. The lever-arm (L1) of the internal moment and the ultimate moment of resistance (U1) follow at lines 1180 and 1190. The required and actual values of the moment of resistance are compared at line 1200 in order to assess whether the section has a sufficient reserve of strength. The conclusion which is drawn at line 1210 dictates the content of the output.

3.3.11 Examples of Output from the Prestressed Concrete Design Library Sub-programs PB4, PB2, PB3 and PB5

A typical set of solutions given by these sub-programs appears below. The design is that for a simply supported prestressed concrete beam spanning 15.2 m and carrying an imposed service load of 10.4 kN/m.

PB4 Run: Identical strings of non-zero 'composite' and prestressed section dimensions reflect the fact that this design is for a normal prestressed beam. The first design attempt resulted in a message to increase the section dimensions. The designer's response to this was to increase the depth, but to keep the remaining dimensions the same; hence the form of the second section input. This section yielded a solution which was accepted.

PB2 Run: The PB4 output supplied the data for this program run. Due to the relatively narrow web only one of the eleven tendons considered by the program was found to be suitable.

PB3 Run: The output from PB4 and PB2 supplied the data for this program run. The limitations of web and flange width confined the choice of tendon group to three rows of single tendons.

PB5 Run: The output from PB4 and PB3 supplied the data for this program run. The proposed section proved to have a satisfactory reserve of strength at ultimate load.

```
PB4

YOUR VALUES OF SPAN, SUPERLOAD, C, CT AND LOSSFACTOR ARE
 ? 15.2, 10.4, 16.67, 0, 0.8

COMPOSITE SECTION DIMENSIONS ARE
 ? 800, 500, 200, 200, 225, 150, 175, 150, 0, 0
```

PRESTRESSED SECTION DIMENSIONS ARE
? 800,500,200,200,225,150,175,150

INCREASE SECTION DIMENSIONS

COMPOSITE SECTION DIMENSIONS ARE
? 900,1,1,1,1,1,1,1,1,1

PRESTRESSED SECTION DIMENSIONS ARE
? 900,1,1,1,1,1,1,1

SOLUTION POSSIBLE

IF SOLUTION REQUIRED TYPE 1 ELSE TYPE 0 ? 1

LIMITING ECCENTRICITIES

E1= 81.2926 MM E2= 358.71 MM

Y2= 553.736 MM

E3= 94.9407 MM E4= 333.237 MM

R2= 7.10218E-3

X4= 4.00479E+6

REQUIRED ULTIMATE MOMENT = 691.618 KNM

IF SECTION PROFILE IS REQUIRED THEN TYPE 1 ELSE 0 ? 1

```
    *************************
    *************************
    *************************
    *************************
    *************************
    *************************
            *******
            *******
            *******
            *******
            *******
            *******
            *******
            *******
            *******
            *******
            *******
            *******
            *******
            *******
            *******
            *******
            *******
            *******
            *******
          **********
          **********
          **********
          **********
          **********
```

COMPOSITE SECTION DIMENSIONS ARE
? 0,0,0,0,0,0,0,0,0,0

RUNNING TIME: 3.8 SECS   I/O TIME :   4.9 SECS

PB2

WEB THICKNESS OF SECTION (MM)= ? 150

LOWER LIMITING ECCENTRICITY E1 (MM)= ? 81.29

ESTIMATED ECCENTRICITY OF PRESTRESSING FORCE (MM) = ? 358.7
VALUES OF Y2,R2 AND X4 TAKEN FROM
PROGRAMME PB4 OUTPUT ARE ? 553.7,7.10218E-3,4.00479E+6

PRESTRESSING FORCE FOR ESTIMATED ECCENTRICITY= 1128.89 KN

OUT OF THE 11 STRANDED CABLES CONSIDERED
THE FOLLOWING CABLES CAN BE ACCOMMODATED IN THE WEB ********

CABLE REF. NO.	NO. OF WIRES/ STRAND	NO. OF STRANDS/ CABLE	NOMINAL STRAND DIA.(MM)
1	7	4	12.5

POSSIBLE CABLE GROUPS ************

CABLE REF. NO.	ACTUAL PRES. FORCE(KN)	%AGE EXCESS	NO. OF STEEL CABLES REQD.	SERVICE STRESS	DIST. FROM BTM. FLANGE TO C.G. OF CABLE SYSTEM (MM)
1	1205.12	7	3	1066.1	226.597

RUNNING TIME: 1.7 SECS   I/O TIME :   2.2 SECS

PB3

PROPOSED CABLE - NUMBER OF STRANDS/CABLE
AND STRAND DIAMETER (MM) ARE ? 4,12.5

BOTTOM FLANGE WIDTH (MM)= ? 200
WEB THICKNESS (MM)= ? 150

MAXIMUM NUMBER OF CABLES ACROSS FLANGE= 1
MAXIMUM NUMBER OF CABLES ACROSS WEB= 1

DECIDE UPON CABLE ARRANGEMENT ***********

HOW MANY ROWS ? 3
DISTANCES (MM) FROM BOTTOM OF BEAM

TO C.G. OF FIRST ROW AND C.G. OF CABLE GROUP ARE
? 100,226.6
NUMBER IN ROW 1   ? 1
NUMBER IN ROW 2   ? 1
NUMBER IN ROW 3   ? 1

DISTANCE TO CENTRE OF FIRST ROW IS 100 MM

DISTANCE BETWEEN CENTRES OF ROWS IS 126.6 MM

MINIMUM END BLOCK WIDTH TO ACCOMMODATE
ANCHORAGES= 280 MM

IF ALL ARRANGEMENTS HAVE BEEN CONSIDERED
THEN TYPE 1 ELSE 0 ? 1

RUNNING TIME: 1.2 SECS   I/O TIME :   3.1 SECS

PB5

REQUIRED MULT (KNM)= ? 691.62
YOUR VALUES OF OVERALL BEAM DEPTH,
UPPER FLANGE WIDTH,
OUTER AND INNER FLANGE THICKNESSES
AND WEB THICKNESS (MM) ARE
? 900,500,200,225,150
CHARACTERISTIC CONCRETE STRENGTH (N/MM↑2)
AND LOSS FACTOR ARE ? 50,0.8
NUMBER OF CABLES= ? 3
CABLE TYPE - NUMBER OF STRANDS/CABLE
AND STRAND DIAMETER (MM) ARE ? 4,12.5
HOW MANY ROWS ? 3
NUMBER OF CABLES IN ROW 1   ? 1
NUMBER OF CABLES IN ROW 2   ? 1
NUMBER OF CABLES IN ROW 3   ? 1
DISTANCE (MM) FROM BOTTOM OF BEAM
TO C.G. OF FIRST ROW= ? 100
DISTANCE (MM) BETWEEN ROWS= ? 125

ACTUAL MULT GREATER THAN REQUIRED

ULTIMATE MOMENT OF RESISTANCE= 1009.76 KNM

DEPTH OF NEUTRAL AXIS= 171.195 MM

STEEL STRESS AT ULTIMATE LOAD EXPRESSED AS A
PERCENTAGE OF THE DESIGN STRENGTH

AT ROW 1   99 %
AT ROW 2   99 %
AT ROW 3   98 %

RUNNING TIME: 2.4 SECS   I/O TIME :   4.3 SECS

Chapter 4

The Computer Aided Design of Rigid Frames—Aspects Common to All Frames

4.1 Introduction

The design of a member from an indeterminate structure is not significantly different from that of its counterpart in a determinate structure; the design requirements for both are essentially the same, but in the former case they are more difficult to assess because the very proportions of each member affect in some degree the behaviour of every other member in the structure. In a structure comprising a large number of members the practical effects of changing the section proportions of a single member are fortunately local—but the effects are often significant within the sphere of influence of the modified member. This interdependence of parts poses a problem common to both the manual and computer aided design of continuous frames alike—that of initially assessing the relative second moments of area of the members comprising the structure. If precisely correct values can be assigned initially then the structure is effectively designed before a formal solution is even begun. But this is a counsel of perfection. In practice rigid frame design becomes an iterative process; one from which an acceptable solution will always be forthcoming as the outcome of a cycle of design iterations in which each iteration produces an increasingly better approximation.

The manner in which starting values for second moments of area are assessed is influenced by the type of design calculation in which they are to be used. For an automatic iterative calculation it is almost sufficient to say that any initial approximation is better than none at all. Indeed, it will be shown in Chapter 6 that in the case of designs involving the choice of steel sections from a standard list an automatic solution will always be possible and the initial values assigned to the second moments of area are largely immaterial to the quality of the final result. If for reinforced concrete structures a different set of constraints is accepted from those implied by a list of standard sections, e.g. designer specified rib and column proportions and steel ratios, then these too will furnish sufficient data to lead to a successful automatic design conclusion. In this case the starting values for the second moments of area are conveniently based upon sections which are derived from maximum allowable deflection and slenderness criteria. In comparison with steel structures, however, it must be expected that concrete structures will require a larger number of design iterations to attain a solution because of the significant contribution that the (initially unknown) structural self-weight makes towards the total loading.

In contrast with this, decision design based programs require informed initial guesses to be made in order to reasonably restrict the overall design time. In the case of continuous reinforced concrete frames this is not such a confining requirement when it is appreciated that the concrete dimensions (which to a degree affect the distribution of force actions) themselves represent a wide range of strengths (resistance to bending, shear, axial load and torsion) depending upon the type and quantity of reinforcement incorporated. During the design of such structures it often happens that a set of guessed section dimensions does not, on analysis of the structure, produce the force actions for which the sections were initially proportioned; but because of their inherent strength range they may still be reinforced to sustain the new force actions. In both automatic design and decision design it is the finite step in strength between one practical section and the next that allows convergence to a unique solution.

It must be accepted that structural analysis is an important aspect of the overall design process and that it is more necessary to choose a method which reflects the behaviour of the structural material than it is to make a choice which is based purely on its suitability for computer programming and operation. Whilst stiffness matrix analysis lends itself to compact programming procedures and efficient computer solution most available matrix analysis programs are based upon assumed linear behaviour. Their results are only directly applicable to a material such as steel working within its elastic range. In the case of reinforced concrete, its non-linear stress/strain characteristics, its creep, shrinkage, cracked regions and non-uniform distribution of reinforcement all conspire to produce deformations which are at times grossly at variance with those predicted by linear elastic theory. When dealing with such materials it might be wiser either to favour a less sophisticated analytical method (albeit one which is probably based upon assumed elastic behaviour) and to temper the more obvious discrepancies between theoretical and actual behaviour with engineering judgement, or to opt for even greater sophistication than that offered by linear matrix analysis by taking account of non-linear behaviour.

Anyone able to specify the parameters of a design may use an automatic design based program successfully, but only competent designers will be capable of producing reasonable results from a decision design based program. This follows from the fact that in decision design the computer is essentially nothing more than an aid to calculation; it extends the designer's powers but abrogates few of his responsibilities.

4.2 Describing the Structure and its Loading to the Computer

4.2.1 Regular Rectangular Frames

Multi-storey, multi-bay frames have a regular geometry and distribution of joints which make the assignment of structural information to the computer,

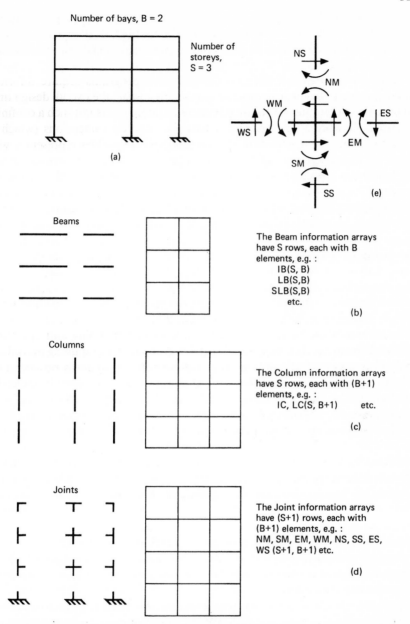

Number of bays, B = 2

Number of storeys, S = 3

(a)

(e)

NS NM WM WS ES EM SM SS

Beams

The Beam information arrays have S rows, each with B elements, e.g. :
IB(S, B)
LB(S,B)
SLB(S,B)
etc.

(b)

Columns

The Column information arrays have S rows, each with (B+1) elements, e.g. :
IC, LC(S, B+1)   etc.

(c)

Joints

The Joint information arrays have (S+1) rows, each with (B+1) elements, e.g. :
NM, SM, EM, WM, NS, SS, ES, WS (S+1, B+1) etc.

(d)

Figure 4.1 Arrays associated with the definition of multi-bay, multi-storey frames

and its subsequent handling and transfer, a relatively easy task. Each member, joint or force action—or indeed any quantity which may be used to describe the structure or its behaviour—can be represented by the content of an element in a rectangular array. Moreover, the array element and the quantity which it represents occupy the same relative positions in the array and the structure respectively. This one-to-one relationship between the structure and its associated arrays greatly simplifies the identification of numerical information with its origin in the structure.

Neither joints nor members need to be formally numbered. Their locations in the frame are known when the integer variables I and J are given values which specify the floor or storey level and the bay or vertical column line at which arithmetical operations are to be carried out.

If the number of storeys in a frame is denoted by S, and the number of bays by B, then the structure will contain $S*B$ beams. For the frame shown in Fig. 4.1a, $S*B = 6$. Some information pertaining to these beams may be conveniently held in 6-element arrays comprising 3 rows and 2 columns. Thus SLB( ), IB( ) and LB( )—see Fig. 4.1b—could be arrays which hold information concerning the beam superimposed loads, second moments of area and spans, respectively. (It should be noted that in all but the simplest programming languages the variables and array names may take more complex descriptive titles than are allowed in BASIC, a facility which greatly simplifies the understanding of the printed program.)

If the integer variables I and J are each made equal to 2 say, to denote that we are interested in the array element defined by the intersection of the second row and the second column, then the content of LB(2, 2) will be the value of the span at the second floor level in the right hand bay. More frequently we would need to refer to the general element LB(I, J) and the other quantities associated with it such as IB(I, J) and SLB(I, J).

The frame also has $(S + 1)*(B + 1)$ joints and $S*(B + 1)$ columns. Associated joint arrays will therefore have $(S + 1)$ rows and $(B + 1)$ columns whilst the column member arrays will have S rows and $(B + 1)$ columns—see Figs. 4.1c and d.

Using the convention that at a four-member joint the ends of the members are labelled North, South, East and West (see Fig. 4.1e), it is convenient to identify the bending moments at the ends of the members by NM, SM, EM and WM. In a similar way the shear forces which act at the ends of members are called NS, SS, ES and WS. Each family of force actions is stored in its own array which has $(S + 1)$ rows and $(B + 1)$ columns (see Fig. 4.1d). The force arrays will necessarily contain a number of permanent zero values. For example, all the elements in the first row of NM( ) will be zero because there are no frame columns above roof level.

In order to make calculations which involve the elements from two or more arrays it is necessary to establish relationships between the positions of elements in those arrays. This is most readily done by inspection. Referring to Figs. 4.1b, c and d it can be seen that the superimposed load on the beam identified by LB(I, J) is SLB(I, J). The bending moment at the left hand end of this member is EM(I, J) and at the right hand end, WM(I, J + 1). NM(I, J) is the bending moment at the lower end of the column member whose length is LC(I − 1, J).

By calculating the shear force at the end of a beam as a function of the beam span, the load upon it and the support bending moments the reader may verify the following relationship for $1 \leqq I \leq S$ and $1 < J \leqq (B + 1)$:

$$WS(I, J) = SLB(I, J - 1) * LB(I, J - 1)/2 - (EM(I, J - 1) + WM(I, J))/LB(I, J - 1)$$

4.2.2 Irregular Frames

In contrast with the approach which was discussed in Section 4.2.1 a more general method of assigning structural information to the computer may be used. This method, which is suitable for both geometrically regular and irregular frames alike, is based upon a joint numbering system; its main application is in conjunction with matrix methods of analysis. Whilst in theory the order in which joints are numbered should not matter, in practice the numbering system has a significant effect on the size of the stiffness matrix array. An effort should therefore be made to number the joints in such a way that the maximum numerical difference between the two joints at the ends of any member is kept to a minimum.

An irregular frame is shown in Fig. 4.2a. It comprises 8 joints and 7 members; member number (3) for example is connected to joint numbers (3) and (4)—the direction in which the member is to be considered is indicated by an arrow.

Three arrays of data inform the computer of the frame's geometry, member properties and support conditions. These arrays, called JC( ), MP( ) and SC( ), are shown in Figs. 4.2b, c, and d. For a structure with NJ joints, array JC( ) has NJ rows and 3 columns. The first column records joint reference numbers in their correct order. The remaining two columns hold the associated x and y joint coordinates—in this case taken from joint number (1) as datum.

For a structure comprising NM members the second array, called MP( ), has NM rows and 5 columns. The member reference numbers are held in the first column of this array in their correct order. The remaining elements in a row contain the two joint reference numbers which define the ends of the member (entered in the order defined by the member direction), the cross-sectional area of the member and its second moment of area. For the initial analysis of an iterative design solution it is probable that A and I would be assigned programmed values. In this, two options are possible: either A and I are related to actual sections or the cross-sectional areas are given artificially inflated values which would have the effect of virtually suppressing the secondary force actions due to axial strains. This means that the initial second moments of area could be treated as relative rather than absolute values. Following the first round of analysis and section selection more realistic assessments of A and I would then be available for subsequent analyses.

With these two arrays the computer has information concerning the location, orientation and properties of all the members in the structure. One instance in which information from these arrays must be related occurs when the member

114

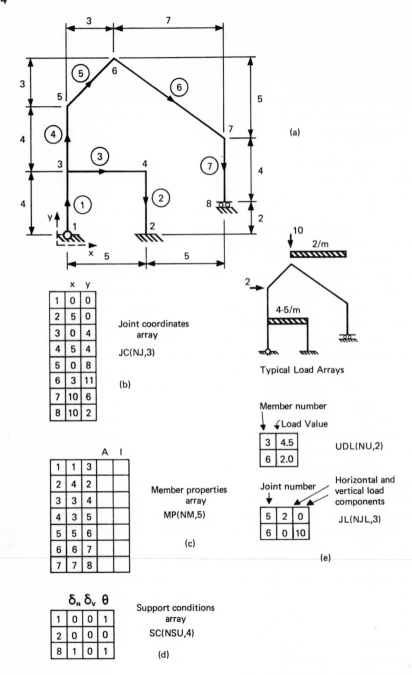

Figure 4.2 Arrays associated with the definition of irregular frames

lengths are calculated. Referring to array MP( ), Fig. 4.2c, if a member reference number is defined by I then the joint reference number at each end of that member will be MP(I, 2) and MP(I, 3). From array JC( )—see Fig. 4.2b—the *x* and *y* coordinates of joint number MP(I, 2) are therefore JC(MP(I, 2), 2) and JC(MP(I, 2), 3) respectively. Similarly the coordinates of the other joint are JC(MP(I, 3), 2) and JC(MP(I, 3), 3). Note here that the *row* location of an element in one array (i.e. JC( )) is defined by the *content* of an element from another array (i.e. MP( )). Using these coordinates to define the ends of a member its length may therefore be calculated.

To complete the structural description a third array contains the support conditions. In Fig. 4.2d this array is called SC( ). The first column of the array holds the reference numbers of joints which constitute supports. The type of support involved will impose particular zero deformations; for example the support at joint number (1) is hinged and the deformations in both the *x* and *y* directions (δ_H and δ_V) will therefore be zero. These values are recorded in the second and third elements of the row. The numeral 1 in the fourth element of that row tells the computer that the support rotation is unknown. If NSU is the number of supports then array SC( ) will have NSU rows and 4 columns.

Further arrays are required to store the loading data. In this respect there will be as many arrays as there are basic load types. Two such arrays are shown in Fig. 4.2e where UDL( ) and JL( ) are concerned with transverse uniformly distributed loads and joint point loads respectively.

4.3 Structural Analysis

4.3.1 Introduction

The principles of structural behaviour which were established by 19th century applied mathematicians still form the basis of many contemporary methods of structural analysis. A direct application of these principles describes structural behaviour in a set of linear algebraic simultaneous equations which relate the structural geometry and displacements to the applied loads. For long after the principles were evolved the weight of arithmetic involved in the solution of the equations precluded their application from anything but simple structural forms. With the advent of electronic digital computers and their proficiency in executing repetitive arithmetic it was natural for analysts to reappraise those early ideas and to develop them in more sophisticated forms. And considering the nature of the problem it was inevitable that matrix algebra would be found to be a powerful shorthand with which to describe the overall behaviour of a structure and the interdependence of its parts. The strength of matrix methods of structural analysis lies with the fact that their use make it no longer necessary to catalogue structures into precise types or to apply a special method of analysis to a particular structural type. All structures belonging to the same category, however loaded, yield to the same treatment.

The traditional design office methods of slope-deflection and moment distribution, whilst stemming from the same roots as matrix methods, are only easily applied in the computer context to structures comprising a regular combination of rectangular cells. Even so, a large number of structures take this form and since moment distribution in particular (or preferably one of its derivatives, successive shear corrections) often leads to a more rapid solution and generally requires less computer storage space than do matrix methods, it can often be used to advantage.

The function of analysis is to furnish an assessment of the force actions and deformations induced by a load system. The necessarily iterative nature of the design of indeterminate structures means that throughout the initial design stages the structural properties are in a state of flux; accurate assessments of force actions are therefore only accurate in so far as they relate to the immediately previous assessment of section proportions. Whilst, because of program length, it is desirable to incorporate only one method of analysis in a design program a case can be made for the savings in computing time to be gained by arranging for successive analyses to produce, from a relatively coarse beginning, increasingly accurate results. With a relaxation type of solution this may be done by aranging for the criterion on which an analysis is terminated to become more severe as the number of overall design iterations increases. In the case of a stiffness matrix approach it might be profitable to reduce the number of equations to be solved by suppressing the secondary effects due to axial deformation during the initial design stages.

The expected relationship between the calculated force actions and those actually attained in practice should affect the choice of analytical method and the degree of accuracy to which it is pursued. Material behaviour, construction tolerances and the precision with which the loads have been estimated will also have been considered when making this decision. For structures deserving an 'accurate' treatment a two-level investigation may save computing time. At the lower level, design programs which are based upon acceptable approximate methods of analysis could be used to produce a number of rapid, but relatively unrefined solutions to the same problem for qualitative comparison; Program RCF12, Chapter 5, could be considered in this way. On the assumption that the relative merits of each solution were still valid even after refinement only one need then be chosen for a more thorough investigation.

In keeping with the general approach so far established of describing *familiar* problems in programming terms, the structural analysis programming which is described in the following sections is confined to a discussion of the slope-deflection method. It will be seen that this approach has much in common with stiffness matrix methods and should serve as a useful preliminary to further study.

4.3.2 An Outline of the Slope-Deflection Method

The slope-deflection equations:

$$M_{AB} = \frac{2EI}{L}\left(2\theta_A + \theta_B - \frac{3(\Delta_B - \Delta_A)}{L}\right) - M_{AB}^F$$

$$M_{BA} = \frac{2EI}{L}\left(\theta_A + 2\theta_B - \frac{3(\Delta_B - \Delta_A)}{L}\right) + M_{BA}^F$$

relate the bending moments generated at the ends of a member AB (see Fig. 4.3) to the rotations of its ends (θ_A and θ_B), the rotation of the member ($\Delta_B - \Delta_A$)/L and the fixed end moments (M_{AB}^F and M_{BA}^F) due to transverse loading.

In a frame of the type shown in Fig. 4.4a the pattern of members and supports prevent vertical sway. If, therefore, all axial member deformations are neglected then the *beams* will not rotate. Writing $K = EI/L$ the *beam* slope-deflection equations become:

$$M_{AB} = 4K^B\theta_A + 2K^B\theta_B - M_{AB}^F$$
$$M_{BA} = 2K^B\theta_A + 4K^B\theta_B + M_{BA}^F \qquad (4.1)$$

If the columns are not subjected to transverse loads within their lengths then their slope-deflection equations will not include a fixed end moment term; but because horizontal sway is possible the *columns* may rotate. For *columns* the equations become:

$$M_{AB} = 4K^C\theta_A + 2K^C\theta_B - \frac{6K^C\Delta_B}{L^C} + \frac{6K^C\Delta_A}{L^C}$$

$$M_{BA} = 2K^C\theta_A + 4K^C\theta_B - \frac{6K^C\Delta_B}{L^C} + \frac{6K^C\Delta_A}{L^C} \qquad (4.2)$$

For a frame with B-bays and S-storeys there will be $S*(B + 1)$ joint rotations (θ) and S sway displacements (Δ). If axial member deformations are neglected then $S*(B + 2)$ equations are sufficient to determine the unknown displacements. These equations are furnished by considering the moment equilibrium of each joint and the horizontal equilibrium at each floor level. Assuming that values of K^B, K^C, L^B, L^C, M_{AB}^F, M_{BA}^F, θ and Δ for the structure are stored in arrays dimensioned as shown in Fig. 4.4b, then the moment equilibrium equation for a joint specified by its (I, J) position in the structure may be written as:

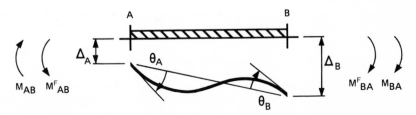

Figure 4.3 Bending moments generated at the ends of member AB

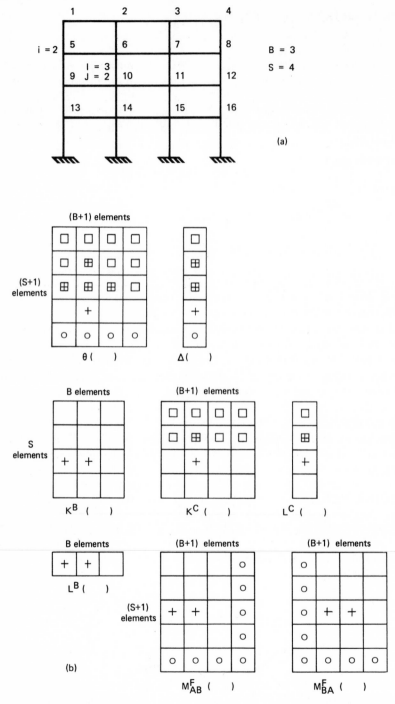

Figure 4.4 Arrays associated with the equilibrium equations

$$4K^B_{I,J}\theta_{I,J} + 2K^B_{I,J}\theta_{I,J+1} + 4K^B_{I,J-1}\theta_{I,J} + 2K^B_{I,J-1}\theta_{I,J-1} + 4K^C_{I-1,J}\theta_{I,J}$$

$$+ 2K^C_{I-1,J}\theta_{I-1,J} - 6K^C_{I-1,J}(\Delta_{I-1} - \Delta_I)/L^C_{I-1} + 4K^C_{I,J}\theta_{I,J} + 2K^C_{I,J}\theta_{I+1,J}$$

$$- 6K^C_{I,J}(\Delta_I - \Delta_{I+1})/L^C_I = -(M^F_{AB(I,J)} - M^F_{BA(I,J-1)}) \tag{4.3}$$

This states that the sum of the moments at the ends of members meeting at a joint is equal and opposite to the out-of-balance moment acting at that joint.

The equation which defines horizontal equilibrium at a floor level which is identified by i is:

$$\sum_{J=1}^{J=B+1}\left(\frac{6K^C_{i-1,J}}{L^C_{i-1}}\left(\theta_{i-1,J} + \theta_{i,J} - \frac{2}{L^C_{i-1}}(\Delta_{i-1} - \Delta_i)\right)\right.$$

$$\left.- \frac{6K^C_{i,J}}{L^C_i}(\theta_{i,J} + \theta_{i+1,J} - \frac{2}{L^C_i}(\Delta_i - \Delta_{i+1}))\right)$$

$$= \text{The horizontal load applied at floor level} \tag{4.4}$$

The LHS of this equation represents the sum of the column shear forces at a given floor level—each shear force is the result of dividing the sum of the moments acting at the end of a column by its length.

In Fig. 4.4b the array elements involved in equation (4.3) are marked thus +, whilst those in equation (4.4) are marked □.

The 3-bay, 4-storey frame shown in Fig. 4.4a may be analysed by solving the 20 simultaneous linear algebraic equations resulting from 16 joint equilibrium conditions (the column/foundation connections are not considered here because their displacements are known to be zero), and 4 horizontal equilibrium conditions.

When the joints are numbered (along rows) from 1 to 20, with the first joint located at $I = J = 1$, the relationship between a joint reference number and its (I, J) location is:

JOINT REFERENCE NUMBER $= (I - 1)*(B + 1) + J$

If the first 16 equations are set up in the order of their joint reference numbers, then for $(I = 3, J = 2)$—see Fig. 4.4a—the joint reference number (and hence the equation number) will be $(3 - 1)*(3 + 1) + 2 = 10$. Thus LHS of equation number (10) becomes:

$$4(K^B_{3,2} + K^B_{3,1} + K^C_{2,2} + K^C_{3,2})\theta_{10} + 2K^B_{3,2}\theta_{11} + 2K^B_{3,1}\theta_9 + 2K^C_{2,2}\theta_6$$

$$+ 2K^C_{3,2}\theta_{14} - \frac{6K^C_{2,2}\Delta_2}{L^C_2} + 6\left(\frac{K^C_{2,2}}{L^C_2} - \frac{K^C_{3,2}}{L^C_3}\right)\Delta_3 - \frac{6K^C_{3,2}\Delta_4}{L^C_3}$$

The last four equations are set up by considering the floors in order, numbered consecutively from the top of the frame. For $i = 2$ the LHS of equation number (18) takes the form:

120

$$\sum_{J=1}^{J=4} \left(\frac{6K_{1,J}^C \theta_{1,J}}{L_1^C} + \left(\frac{6K_{1,J}^C}{L_1^C} + \frac{6K_{2,J}^C}{L_2^C} \right)\theta_{2,J} - \frac{6K_{2,J}^C \theta_{3,J}}{L_2^C} \right.$$

$$\left. - \frac{12K_{1,J}^C \Delta_1}{(L_1^C)^2} + \left(\frac{12K_{1,J}^C}{(L_1^C)^2} + \frac{12K_{2,J}^C}{(L_2^C)^2} \right)\Delta_2 - \frac{12K_{2,J}^C \Delta_3}{(L_2^C)^2} \right)$$

It will be appreciated that, in effect, the LHS of each equation also contains zero coefficients to match the displacements not involved in that particular equation of equilibrium. The pattern of displacement coefficients for the 20

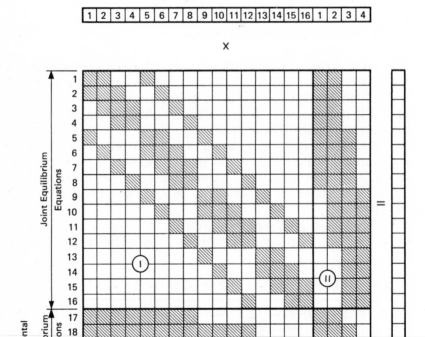

Figure 4.5 Pattern of displacement coefficients

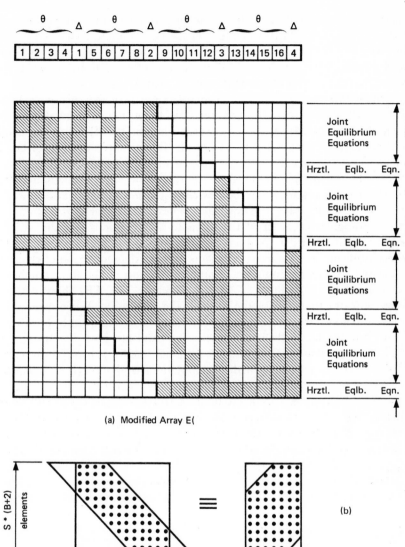

(a) Modified Array E( )

(b)

Figure 4.6 Displacement coefficients concentrated within a diagonal band

equations is shown in array E( ), Fig. 4.5, where zero coefficients are shown as blank elements and non-zero coefficients are hatched. A notable feature of the array is the symmetry of its contents about the leading diagonal. Later in this section it is shown how this property may be exploited.

The content of each non-zero element in the array is a function of member properties. The distribution of displacement coefficients throughout the array reflects the way in which the members of the structure are connected to each other. Array E( ) represents the *Frame Stiffness Matrix* (K) for the particular case of a frame with members of infinitely large cross-sectional area, i.e. it is assumed that the axial displacement between joints is zero.

To pursue this analogy with matrix methods a little further, array D( )—see Fig. 4.5—which represents the unknown displacements, is equivalent to the *Displacement Vector* (d); array P( ), which contains the RHS's of the equilibrium equations, is the *Load Vector* (p). In this example array P( ) has 20 elements. The contents of the first 16 elements are the RHS's of equation type (4.3) and the four remaining elements represent the RHS's of equation type (4.4). The set of equations may be described by the single equation p = Kd. A method of solving this set of equations is described in the next section.

By re-arranging the order of the terms in the equations and the order in which the equations are derived it is possible, whilst still preserving symmetry, to concentrate all the non-zero coefficients within a diagonal band of elements, the 'bandwidth' comprising a decreasing proportion of the total number of displacements as the number of storeys increases. Fig. 4.6a shows such an arrangement of equations for the frame under discussion. When arranged in this way the non-zero coefficients are embraced by a 'bandwidth' of (4B + 7) elements and all the coefficients outwith this band are zero. By storing only those coefficients which fall within the diagonal band it is possible to replace the $(S(B + 2))^2$ element array with one containing $S(4B + 7)$ (B + 2) elements —see Fig. 4.6b. In the case of large multi-storey structures the consequent reduction in the array size needed to hold the stiffness matrix makes substantial savings in the computer storage requirements. A further useful reduction in the storage requirements for this array can be made by recognising that, since the coefficients within the band are symmetrical about the leading diagonal, the content of the elements on and above the leading diagonal wholly reflect the properties of the structure; only those elements therefore need to be stored.

These devices naturally affect the way in which array elements are manipulated during the solution of the equations. In the ensuing discussion the total stiffness matrix is considered and the refinements mentioned above are ignored.

4.3.3 Solving the Equations

A simple and orderly way of solving a set of simultaneous equations is to use the method of successive elimination of variables. The method will be discussed in relation to a set of n simultaneous equations, where $x_1 \ldots\ldots x_n$ are the variables to be determined:

$$x_1 \quad x_2 \quad x_3 \ldots\ldots\ldots \quad x_n$$

$$a_{11} + a_{12} + a_{13} + \ldots\ldots + a_{1n} = C_1 \qquad (1)$$
$$a_{21} + a_{22} + a_{23} + \ldots\ldots + a_{2n} = C_2 \qquad (2)$$

$$a_{31} + a_{32} + a_{33} + \ldots\ldots + a_{3n} = C_3 \qquad (3)$$

$$(4.5)$$

.
.
.
.
.

$$a_{n1} + a_{n2} + a_{n3} + \ldots\ldots + a_{nn} = C_n \qquad (n)$$

The first object is to eliminate x_1 from the second and subsequent equations. This is done in two stages:

1. Assuming that $a_{11}, a_{21}, \ldots\ldots a_{n1} \neq 0$ then each equation is divided by its coefficient of x_1. The effect of this is to make all the non-zero coefficients in the first column equal to 1;
2. The first equation is then subtracted from each equation in turn. Thus the first coefficient in the modified equations $(2) \ldots\ldots (n)$ becomes zero.

The set of equations (4.5) therefore takes the form:

$$x_1 \quad x_2 \quad x_3 \ldots\ldots\ldots \quad x_n$$

$$
\begin{array}{ll}
1 + a'_{12} + a'_{13} + \ldots\ldots + a'_{1n} = C'_1 & (1) \\
0 + a'_{22} + a'_{23} + \ldots\ldots + a'_{2n} = C'_2 & (2) \\
0 + a'_{32} + a'_{33} + \ldots\ldots + a'_{3n} = C'_3 & (3)
\end{array}
$$

$$(4.6)$$

.
.
.
.
.

$$0 + a'_{n2} + a'_{n3} + \ldots\ldots + a'_{nn} = C'_n \qquad (n)$$

where the primes indicate that the coefficients are now modified as a result of the arithmetical operations which have just been carried out. In the set of equations (4.5) equation (1) is called the *pivotal equation* and the first element in that equation is the *pivotal element*.

By applying the same procedure to the equations $(2) \ldots\ldots (n)$ in the modified set (4.6), with equation (2) as the new pivotal equation and a'_{22} as the pivotal element, x_2 is eliminated from the equations (3) to (n). A further $(n-2)$ similar operations carried out on successively smaller blocks of equations converts the original set (4.5) to the form:

$$x_1 \quad x_2 \quad x_3 \ldots\ldots\ldots \quad x_n$$

$$
\begin{array}{ll}
1 + a'_{12} + a'_{13} + \ldots\ldots + a'_{1n} = C'_1 & (1) \\
0 + 1 + a''_{23} + \ldots\ldots + a''_{2n} = C''_2 & (2)
\end{array}
$$

$$0 + 0 \quad + 1 \quad + \ldots \ldots + a'''_{3n} = C'''_3 \qquad (3)$$

$$(4.7)$$

.

.

.

.

$$0 + 0 \quad + 0 \quad + \ldots \ldots + 1 \quad = C_n^{('n)} \qquad (n)$$

Equation (n) of the set (4.7) shows that $x_n = C_n^{('n)}$. The value of the variable x_{n-1} is calculated by substituting the known value of x_n into equation $(n-1)$. The remaining variables are determined in a similar way by substituting the values of known variables into each equation in turn.

Excepting for errors arising from 'rounding-off' calculated quantities the method described is an exact one which readily lends itself to computer programming techniques. The load-displacement equations which describe the behaviour of multi-cell structures are almost invariably well conditioned and can be expected to yield 'accurate' values for the displacements. However, one modification to the standard method, which requires little extra programming effort, can reduce the effect of a possible source of inaccuracy.

According to the method described above the pivotal equation was defined as being the first equation in the block of equations waiting to be processed. But if the pivotal element is numerically small in comparison with the remaining elements of the column which it occupies, then the modified coefficients in the pivotal equation will probably be large in comparison with the quantities from which they will eventually be subtracted. This situation is a potential source of accumulative errors which may be largely rectified by 'exchanging pivots'. Before the process of eliminating a variable is begun the location is found of the numerically largest coefficient in the first column of elements of the remaining block. The equation which is defined by the location of this coefficient is then chosen as the pivotal equation and exchanged with the first equation in this block. By choosing the pivotal equation in this way, its effect on the remaining equations will be minimal.

The basic solution is complete when the displacements have been determined. The moments at the ends of each member are calculated by substituting the relevant loads and displacements into equations (4.1) and (4.2).

4.3.4 Program Specification—SD1

SD1 is a structural analysis program which is based upon the slope-deflection method. Its application is confined to the analysis of multi-bay, multi-storey frames having zero displacements (both rotational and linear) at foundation level. The load-displacement equations are solved by the method of successive elimination of variables. Each analysis is limited to a single load condition which may comprise any chosen pattern of uniformly distributed beam loads and/or horizontal loads applied at floor levels. The results output comprises a list of member end-moments and joint displacements.

Even within the limitations imposed by the single structural type, an expansion of the program area which deals with the calculation of force actions to include the shear and axial forces and the maximum span bending moments would make SD1 a more useful design tool.

4.3.5 The SD1 Flow Diagram

The flow diagram for this program is shown in Fig. 4.7. Because its form is substantially linear the flow diagram could have been replaced by an ordered list of essential program areas. The only major loops in the program are those introduced to enable further problems to be solved during a single program run.

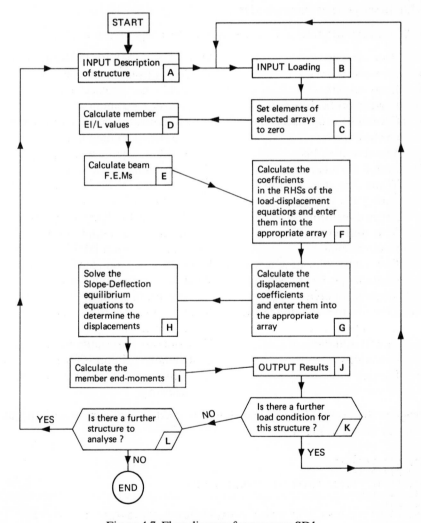

Figure 4.7 Flow diagram for program SD1

The input of structural and loading parameters are treated in separate blocks (A and B) so that further load conditions may be examined without the necessity for repeatedly reading the structural geometry into the computer. At Block (C) certain arrays are formally cleared to ensure that the state of their contents is in no doubt when they come to be used. Blocks (D), (E), (F), (G), (H) and (I) represent the main steps to be followed in an analysis. The calculation of member EI/L values and fixed end-moments at Blocks (D) and (E) clears the way for setting up the load-displacement equations at Blocks (F) and (G). Following the solution of these equations at Block (H) the force actions are determined at Block (I).

4.3.6 Description of Program SD1

A list of the variables and arrays used in this program is given below.

A   Row counter
B   Number of bays
C   Load case trigger
D   Further problem trigger
E( ) Contains the displacement coefficients
E   Young's modulus
F   Frame reference number
G   Temporary store for use during an exchange
H   Array dimension
I   Row counter
J   Column counter
K   Row number locating new pivot
L( ) Contains the beam second moments of area
M( ) Contains the column second moments of area
N   Counts a number of operations
O   Locates a row in E( )

P( ) Initially contains the RHS's of the load-displacement equations, finally the displacements
P   Locates a column in E( )
Q( ) Contains the North moments
Q   Locates a column in E( )
R( ) Contains the East moments
R   Locates a column in E( )
S( ) Contains the South moments
S   Number of storeys
T( ) Contains the West moments
T   Locates a column in E( )
U( ) Contains beam UDL's
V( ) Contains wind loads
W( ) Contains column centres (i.e. beam spans)
X( ) Contains storey heights (i.e. column lengths)
Y( ) Contains beam stiffnesses
Z( ) Contains column stiffnesses
Z   Temporary store for use during an exchange

Program SD1 is listed below and should be read in conjunction with the flow diagram shown in Fig. 4.7.

```
100 DIM E(35,35)
110 DIM P(35)
120 PRINT "INPUT FRAME REFERENCE NUMBER";
130 INPUT F
140 PRINT
150 PRINT "ANALYSIS OF FRAME NUMBER"F"BY SLOPE-DEFLECTION METHOD"
160 PRINT
```

```
170 PRINT "INPUT NUMBER OF BAYS AND STOREYS";
180 INPUT B,S
190 PRINT
200 PRINT "INPUT COLUMN CENTRES - (M)"
210 FOR J=1 TO B
220 INPUT W(J)
230 NEXT J
240 PRINT
250 PRINT "INPUT STOREY HEIGHTS - (M)"
260 FOR I=1 TO S
270 INPUT X(I)
280 NEXT I
290 PRINT
300 PRINT "INPUT BEAM I-VALUES - (MM+4)"
310 FOR I=1 TO S
320 FOR J=1 TO B
330 INPUT L(I,J)
340 NEXT J
350 NEXT I
360 PRINT
370 PRINT "INPUT COLUMN I-VALUES - (MM+4)"
380 FOR I=1 TO S
390 FOR J=1 TO B+1
400 INPUT M(I,J)
410 NEXT J
420 NEXT I
430 PRINT
440 PRINT "INPUT YOUNG'S MODULUS - (N/MM+2)"
450 INPUT E
460 PRINT
470 PRINT "INPUT BEAM UDL'S - (KN/M)"
480 FOR I=1 TO S
490 FOR J=1 TO B
500 INPUT U(I,J)
510 NEXT J
520 NEXTI
530 PRINT
540 PRINT "INPUT WIND LOADS - (KN)"
550 FOR I=1 TO S
560 INPUT V(I)
570 NEXT I
580 PRINT
590 FOR I=1 TO S*(B+2)
600 FOR J=1 TO S*(B+2)
610 LET E(I,J)=0
620 NEXT J
630 NEXT I
640 FOR I=1 TO S*(B+2)
650 LET P(I)=0
660 NEXT I
670 FOR I=1 TO S+1
680 FOR J=1 TO B+1
690 LET O(I,J)=0
700 LET S(I,J)=0
710 LET R(I,J)=0
720 LET T(I,J)=0
730 NEXT J
740 NEXT I
750 FOR I=1 TO S
760 FOR J=1 TO B
770 LET Y(I,J)=E*L(I,J)/(W(J)*1E9)
780 NEXT J
790 NEXT I
800 FOR I=1 TO S
810 FOR J=1 TO B+1
820 LET Z(I,J)=E*M(I,J)/(X(I)*1E9)
```

```
830 NEXT J
840 NEXT I
850 FOR I=1 TO S
860 FOR J=1 TO B
870 LET T(I,J+1)=U(I,J)*W(J)+2/12
880 LET R(I,J)=-T(I,J+1)
890 NEXT J
900 NEXT I
910 FOR I=1 TO S
920 FOR J=1 TO B+1
930 LET O=(I-1)*(B+1)+J
940 LET P(O)=-(T(I,J)+R(I,J))
950 NEXT J
960 NEXT I
970 FOR I=1 TO S
980 LET P(S*(B+1)+I)=V(I)
990 NEXT I
1000 FOR I=1 TO S
1010 FOR J=1 TO B+1
1020 LET O=(I-1)*(B+1)+J
1030 LET P=O+1
1040 LET R=O-1
1050 LET Q=(I-2)*(B+1)+J
1060 LET T=I*(B+1)+J
1070 IF I=1 THEN 1100
1080 LET E(0,0)=E(0,0)+4*Z(I-1,J)
1090 LET E(0,Q)=2*Z(I-1,J)
1100 LET E(0,0)=E(0,0)+4*Z(I,J)
1110 IF I=S THEN 1130
1120 LET E(0,T)=2*Z(I,J)
1130 IF J=B+1 THEN 1170
1140 LET E(0,0)=E(0,0)+4*Y(I,J)
1150 LET E(0,P)=2*Y(I,J)
1160 IF J=1 THEN 1190
1170 LET E(0,0)=E(0,0)+4*Y(I,J-1)
1180 LET E(0,R)=2*Y(I,J-1)
1190 NEXT J
1200 NEXT I
1210 FOR I=1 TO S
1220 FOR J=1 TO B+1
1230 LET O=(I-1)*(B+1)+J
1240 LET P=(B+1)*S+I-1
1250 LET Q=P+1
1260 LET R=O+1
1270 IF I>1 THEN 1300
1280 LET E(0,Q)=-6*Z(I,J)/X(I)
1290 GOTO 1330
1300 LET E(0,Q)=6*Z(I-1,J)/X(I-1)-6*Z(I,J)/X(I)
1310 LET E(0,P)=-6*Z(I-1,J)/X(I-1)
1320 IF I=S THEN 1340
1330 LET E(0,R)=6*Z(I,J)/X(I)
1340 NEXT J
1350 NEXT I
1360 FOR I=1 TO S
1370 FOR J=1 TO B+1
1380 LET O=S*(B+1)+I
1390 LET P=(I-1)*(B+1)+J
1400 LET Q=P+B+1
1410 LET R=P-B-1
1420 LET E(0,P)=-6*Z(I,J)/X(I)
1430 IF I=1 THEN 1470
1440 LET E(0,P)=E(0,P)+6*Z(I-1,J)/X(I-1)
1450 LET E(0,R)=6*Z(I-1,J)/X(I-1)
1460 IF I=S THEN 1480
1470 LET E(0,Q)=-6*Z(I,J)/X(I)
1480 NEXT J
```

```
1490 NEXT I
1500 FOR I=1 TO S
1510 LET O=S*(B+1)+I
1520 IF I=1 THEN 1580
1530 FOR J=1 TO B+1
1540 LET E(0,0)=E(0,0)+12*Z(I,J)/X(I)+2+12*Z(I-1,J)/X(I-1)+2
1550 LET E(0,0-1)=E(0,0-1)-12*Z(I-1,J)/X(I-1)+2
1560 NEXT J
1570 GOTO 1610
1580 FOR J=1 TO B+1
1590 LET E(0,0)=E(0,0)+12*Z(I,J)/X(I)+2
1600 NEXT J
1610 IF I=S THEN 1650
1620 FOR J=1 TO B+1
1630 LET E(0,0+1)=E(0,0+1)-12*Z(I,J)/X(I)+2
1640 NEXT J
1650 NEXT I
1660 LET H=S*(B+2)
1670 FOR N=1 TO H
1680 LET K=0
1690 FOR I=N TO H
1700 IF ABS(E(I,N))<K THEN 1730
1710 LET K=ABS(E(I,N))
1720 LET L=I
1730 NEXT I
1740 FOR J=N TO H
1750 LET G=E(L,J)
1760 LET E(L,J)=E(N,J)
1770 LET E(N,J)=G
1780 NEXT J
1790 LET X=P(L)
1800 LET P(L)=P(N)
1810 LET P(N)=X
1820 FOR I=N TO H
1830 IF E(I,N)=0 THEN 1890
1840 LET Z=E(I,N)
1850 FOR J=N TO H
1860 LET E(I,J)=E(I,J)/Z
1870 NEXT J
1880 LET P(I)=P(I)/Z
1890 NEXT I
1900 IF N=H THEN 1990
1910 FOR I=N+1 TO H
1920 IF E(I,N)=0 THEN 1970
1930 FOR J=N TO H
1940 LET E(I,J)=E(I,J)-E(N,J)
1950 NEXT J
1960 LET P(I)=P(I)-P(N)
1970 NEXT I
1980 NEXT N
1990 FOR A=2 TO H
2000 LET I=H-A+1
2010 FOR J=I+1 TO H
2020 LET P(I)=P(I)-E(I,J)*P(J)
2030 NEXT J
2040 NEXT A
2050 FOR I=1 TO S
2060 FOR J=1 TO B
2070 LET R(I,J)=R(I,J)+4*Y(I,J)*P((I-1)*(B+1)+J)
2080 LET R(I,J)=R(I,J)+2*Y(I,J)*P((I-1)*(B+1)+J+1)
2090 LET T(I,J+1)=T(I,J+1)+2*Y(I,J)*P((I-1)*(B+1)+J)
2100 LET T(I,J+1)=T(I,J+1)+4*Y(I,J)*P((I-1)*(B+1)+J+1)
2110 NEXT J
2120 NEXT I
2130 FOR I=1 TO S
2140 FOR J=1 TO B+1
```

```
2150 IF I=S THEN 2200
2160 LET Q(I+1,J)=4*Z(I,J)*P(I*(B+1)+J)
2170 LET Q(I+1,J)=Q(I+1,J)+2*Z(I,J)*P((I-1)*(B+1)+J)
2180 LET Q(I+1,J)=Q(I+1,J)-6*Z(I,J)*(P(S*(B+1)+I))/X(I)
2181 LET Q(I+1,J)=Q(I+1,J)+6*Z(I,J)*P(S*(B+1)+I+1)/X(I)
2190 GOTO 2230
2200 LET Q(I+1,J)=2*Z(I,J)*P((I-1)*(B+1)+J)
2210 LET Q(I+1,J)=Q(I+1,J)-6*Z(I,J)*P(S*(B+1)+I)/X(I)
2220 GOTO 2260
2230 LET S(I,J)=2*Z(I,J)*P(I*(B+1)+J)+4*Z(I,J)*P((I-1)*(B+1)+J)
2240 LET S(I,J)=S(I,J)-6*Z(I,J)*(P(S*(B+1)+I)-P(S*(B+1)+I+1))/X(I)
2250 GOTO 2270
2260 LET S(I,J)=4*Z(I,J)*P((I-1)*(B+1)+J)-6*Z(I,J)*P(S*(B+1)+I)/X(I)
2270 NEXT J
2280 NEXT I
2290 PRINT
2300 PRINT "MEMBER END MOMENTS (KNM)"
2310 PRINT "(MEMBERS INTERSECT AT JOINTS DEFINED"
2320 PRINT " BY THEIR I,J LOCATIONS.)"
2330 PRINT
2340 PRINT "      I   J      NM          SM          EM          WM"
2350 FOR I=1 TO S+1
2360 FOR J=1 TO B+1
2370 PRINT "JOINT"I;J,Q(I,J),S(I,J),R(I,J),T(I,J)
2380 NEXT J
2390 NEXT I
2400 PRINT
2410 PRINT
2420 PRINT "DISPLACEMENTS"
2430 PRINT
2440 PRINT "JOINT ROTATIONS (RADIANS)"
2450 PRINT "       I   J"
2460 FOR I=1 TO S+1
2470 FOR J=1 TO B+1
2480 IF I=S+1 THEN 2510
2490 PRINT "JOINT"I;J,P((I-1)*(B+1)+J)
2500 GOTO 2520
2510 PRINT "JOINT"I;J," 0"
2520 NEXT J
2530 NEXT I
2540 PRINT
2550 PRINT
2560 PRINT "HORIZONTAL DISPLACEMENTS"
2570 PRINT
2580 PRINT "                         DISPLACEMENT"
2590 FOR I=1 TO S+1
2600 IF I=S+1 THEN 2660
2610 IF I>1 THEN 2640
2620 PRINT "    ROOF LEVEL",P(S*(B+1)+I)*1000"MM"
2630 GOTO 2670
2640 PRINT "    FLOOR"S-I+1"LEVEL",P(S*(B+1)+I)*1000"MM"
2650 GOTO 2670
2660 PRINT "FOUNDATION LEVEL           0 MM"
2670 NEXT I
2680 PRINT
2690 PRINT
2700 PRINT "IF FURTHER LOAD CASES FOR THIS FRAME TYPE 1 ELSE 0";
2710 INPUT C
2720 IF C=1 THEN 460
2730 FOR I=1 TO 10
2740 PRINT
2750 NEXT I
2760 PRINT "IF FURTHER FRAMES TO BE ANALYSED TYPE 1 ELSE 0";
2770 INPUT D
2780 IF D=1 THEN 120
2790 END
```

4.3.6.1 Blocks (A) and (B): Statements 100–580

To increase the capacity of the program for solving simultaneous equations the dimensions of E( ) and P( ), the largest arrays in the program, are arbitrarily set to 35 at lines 100 and 110. In terms of B and S (the number of bays and storeys) the number of equations required for the solution of a given frame is $S*(B + 2)$. If the dimensions of the second largest arrays are not increased beyond that which is automatically allocated to them then for a solution by this program to be possible the expressions $(B + 1) \not> 10$, $(S + 1) \not> 10$ and $S*(B + 2) \not> 35$ set the upper limits to the number of bays and storeys that can be handled in a specific case.

The remainder of Blocks (A) and (B) is wholly concerned with the input of structural parameters and loading. In comparison with the PRINT and INPUT statements which are included in Program RCF12 (Chapter 5) the standard to which SD1 requests information should be considered to be the minimum acceptable.

4.3.6.2 Block (C): Statements 590–740

Block (C) is an essential preliminary to the main calculation. The need to clear an array (i.e. to set the contents of its elements to zero) often arises if the array plays a role in the solution of more than one problem during the same program run. If, for example, quantities are entered into an array according to the statement:

$$96 \text{ LET } C(I, J) = C(I, J) + K$$

then unless $C(I, J)$ is set to zero before the next solution it will retain a non-zero value and cause arithmetical errors in subsequent calculations. With this kind of pitfall in mind the arrays E( ), P( ), Q( ), R( ), S( ) and T( ) are all cleared.

4.3.6.3 Block (D): Statements 750–840

By calculating and storing the $K = EI/L$ value for each member at this stage the form of the statements which govern the calculation of the displacement coefficients at Block (G) is simplified. The beam span or column length used to determine K is inferred from the column centres or storey heights which were read into arrays L( ) and M( ) at Block (A).

4.3.6.4 Block (E): Statements 850–900

The arrays R( ) and T( ) store the bending moments which are induced at the left and right hand supports respectively. At this stage in the calculation they are the fixed end moments due to uniformly distributed span loads—see lines 850 to 900. Subsequently, at Block (I), these arrays record the final support moments after account has been taken of the effects of sway and joint rotation.

4.3.6.5 Block (F): Statements 910–990

Array P( ) has a dual role. At the beginning of an analysis its elements represent the RHS's of the load-displacement equations (the load vector); finally it contains the displacements (the displacement vector).

The load vector elements are defined by the RHS's of equations (4.3) and (4.4). The first $S*(B+1)$ elements of P( ) contain $-1*$(the out of balance moment) at each joint. This quantity, calculated at line 940, is related to the correct element in P( ) by recognizing that a joint reference number and the number of its equilibrium equation are identical. If the location of a joint in the frame is specified by its (I, J) values, then the reference number of that joint is $O = (I - 1)*(B + 1) + J$ (see line 930). Thus P(O) is the location of the correct load vector element for joint number 0.

The remaining S elements of P( ) accommodate the RHS's of the horizontal shear equilibrium equations, i.e. the contents of array V( ). By assigning a value of I to each floor level in turn (beginning with $I = 1$ at roof level) the reference numbers for this last group of equations become $S*(B+1) + I$ (see line 980).

4.3.6.6 Block (G): Statements 1000–1650

All the elements in array E( ) were set to zero at Block (C). The purpose of Block (G) is to calculate the non-zero displacement coefficients, as defined by equations (4.3) and (4.4), and to assign them to their correct locations in E( ). Array E( ) can then be thought of either as representing the displacement coefficients in the equations of equilibrium or the stiffness matrix for the frame.

The horizontal and vertical lines which partition array E( ) in Fig. 4.5 delineate four regions, each of which contains a distinctive pattern of coefficients. In the program a separate set of statements is used to insert the appropriate coefficients into each region.

These are: I Statements 1000 to 1200
                 II Statements 1210 to 1350
               III Statements 1360 to 1490
   and IV Statements 1500 to 1650

The form taken by each of these sets of statements follows naturally from the structure of the relevant equation of equilibrium. The statements which govern the pattern of coefficients in Region I will be considered in detail. It is left to the reader to confirm that the statements governing the remaining regions are valid.

Region I will always have as many rows and columns as there are 'free' joints in the frame. We are therefore concerned with $S*(B+1)$ equations of joint equilibrium and the same number of rotational displacements. Equation (4.3) shows that, in terms of the joint rotations, equilibrium is a function of the rotation $(\theta_{I,J})$ of the nominated joint and the rotations $(\theta_{I,J+1}, \theta_{I,J-1}, \theta_{I+1,J}$

and $\theta_{I-1,J}$) of the joints at the remote ends of the connected members. Thus, in general, a joint equilibrium equation will include five non-zero coefficients of θ. The reference number of an equation within Region I, depending as it does on the (I, J) location of the joint, is defined by $O = (I - 1)*(B + 1) + J$ at line 1020. O is also the location of $\theta_{I,J}$ in the displacement vector. Hence E(O, O) must contain the coefficient of $\theta_{I,J}$. P, R, Q and T, calculated at lines 1030 to 1060, are the displacement vector locations (and therefore the column locations in array E( )) of $\theta_{I,J+1}$, $\theta_{I,J-1}$, $\theta_{I-1,J}$ and $\theta_{I+1,J}$ respectively. The elements E(O, P), E(O, R), E(O, Q) and E(O, T) therefore contain the coefficients of these rotational displacements. It follows from equation (4.3) that at an internal joint which connects four members: $E(O, O) = 4*$(the sum of the K-values of members connected by the joint) and E(O, P), E(O, R), E(O, Q) and E(O, T) $= 2*$(the K-value of the relevant connected member), where K is stored appropriately in either Y( ) or Z( ) according to whether the member is a beam or a column.

For each joint in succession the contents of elements E(O, O), E(O, P), etc. are calculated at lines 1080 to 1180. The conditional statements at lines 1070, 1130 and 1160 take account of the fact that peripheral joints connect fewer than four members. The statement at line 1110 recognizes that since full fixity is assumed at foundation level, E(O, T) does not exist when I = S.

4.3.6.7 Block (H): Statements 1660–2040

The load-displacement equations which were set up at Blocks (F) and (G) are solved at this stage in the program. The sequence of operations which reduces the variables by one is identical whether we are concerned with the 1st or the nth variable. But each time the variables are reduced in number the region of array E( ) over which operations take place is diminished by one row and one column—compare the sets of equations (4.5) and (4.6). If there is a total of H rows and columns in array E( ), where $H = S*(B + 2)$, then the reduction operation must be carried out H times. This is the reason why the reduction operation is embraced by the FOR and NEXT statements at lines 1670 and 1980. And whilst N acts as a counter which controls the number of operations, it also controls the region over which these operations are executed. See, for example, lines 1690 and 1740 where, as N increases, the number of rows and columns concerned decreases. Thus, having defined a region, the three following operations are executed within it:

1. To establish the location of the row which has the numerically largest first element (Statements 1680 to 1730);
2. To exchange the contents of this row with the first one in the region—and vice versa (Statements 1740 to 1810);
3. To reduce the number of variables by one (Statements 1820 to 1970).

The first of these operations compares in turn the contents of the first element in each row (E(I, N)) with the contents of K, which was initially set to zero at line 1680. If $E(I, N)$ is numerically greater than the current value of K then its value is registered in K at line 1710. At the same time the row location (I) is stored in the variable (L) at line 1720.

In operation (2) the contents of the two rows can only be exchanged if a further variable is provided in which to store temporarily one of the mobile quantities. At lines 1750 to 1770 the content of $E(L, J)$ is transferred to G, the content of $E(N, J)$ is transferred to $E(L, J)$ and finally the content of G is copied into $E(N, J)$. By executing this operation at each element in the row an exchange of rows is effected. A similar operation on the relevant load vector elements $P(L)$ and $P(N)$ completes the transfer of equations.

The final operation (3) is carried out by dividing both sides of each equation by its own first coefficient $(E(I, N))$—see lines 1820 to 1890—and then subtracting each equation from the first one, an element at a time, at lines 1910 to 1970.

By executing the above operations H times array $E(\ )$ is converted to a form in which the leading diagonal elements are unity and all elements below the leading diagonal are zero. The original contents of $P(\ )$ have also changed. Referring to the set of equations (4.7) we see that the last equation is in the form:

$$1 * x_n = C_n^{('n)}$$

where, in terms of the frame analysis problem, x_n represents the horizontal displacement at 1st floor level and $C_n^{('n)}$ is equivalent to $P(H)$. Thus the last displacement is now known to equal $P(H)$. The penultimate equation is in terms of the variables x_n and x_{n-1} only, and since x_n is known the variable x_{n-1} can be determined. In general an equation now contains only one more variable than the equation that comes after it; all the variables may therefore be determined by considering the equations in their reverse order and substituting known displacement values in their turn (see lines 1990 to 2040).

4.3.6.8 Block (I): Statements 2050–2280

At this stage array $P(\ )$ now holds the displacements. They are used to calculate the beam support moments at lines 2050 to 2120 and the column moments at lines 2130 to 2280. These calculations are based upon equations (4.1) and (4.2).

4.3.6.9 Block (J): Statements 2290–2690

This block is wholly concerned with the output of results.

4.3.6.10 Blocks (K) and (L): Statements 2700–2790

These statements allow the opportunity of investigating further load cases for the current frame, or further frames, during the same program run.

4.3.7 Example of Program SD1 Output

A typical solution given by this program, together with an illustration of the frame which is analysed, appears at the end of this section. The actual member dimensions, I values and loading should be compared with the program data to verify the order in which they are input. It will be seen that each floor level and storey is considered in turn from top to bottom, and that at each level the appropriate items are input from left to right.

The frame analysis follows immediately on completion of the data input. The output results comprise a summary of the bending moments and displacements.

The scope of the program may be extended to that of handling continuous beam and sub-frame analyses by the simple expedient of introducing fictitious members which simulate the prescribed boundary conditions. For example, continuous beam supports can be simulated by assigning a nominal length and infinitesimal I-value to the columns of a single-storey, multi-bay frame. And the fully-fixed condition which is assumed to occur at the ends of members at the boundaries of a sub-frame is managed by introducing fictitious members of massive I-value with which the 'fixed' ended members connect. In either case the rule is to introduce as many fictitious members as will allow the actual structure to be defined as a multi-bay, multi-storey frame, and to assign to those fictitious members an I-value appropriate to the required support condition.

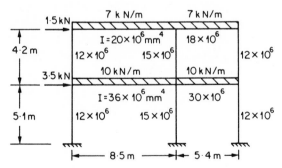

```
SD1

INPUT FRAME REFERENCE NUMBER ? 1

ANALYSIS OF FRAME NUMBER 1 BY SLOPE-DEFLECTION METHOD

INPUT NUMBER OF BAYS AND STOREYS ? 2,2

INPUT COLUMN CENTRES - (M)
 ? 8.5
 ? 5.4

INPUT STOREY HEIGHTS - (M)
 ? 4.2
 ? 5.1
```

```
INPUT BEAM I-VALUES - (MM↑4)
? 20000000
? 18000000
? 36000000
? 30000000

INPUT COLUMN I-VALUES - (MM↑4)
? 12000000
? 15000000
? 12000000
? 12000000
? 15000000
? 12000000

INPUT YOUNG'S MODULUS - (N/MM↑2)
? 200000

INPUT BEAM UDL'S - (KN/M)
? 7
? 7
? 10
? 10

INPUT WIND LOADS - (KN)
? 1.5
? 3.5
```

```
MEMBER END MOMENTS (KNM)
(MEMBERS INTERSECT AT JOINTS DEFINED
 BY THEIR I,J LOCATIONS.)
```

	I	J	NM	SM	EM	WM
JOINT	1	1	0	26.1938	-26.1938	0
JOINT	1	2	0	-19.553	-26.4353	45.9883
JOINT	1	3	0	-9.871	0	9.871
JOINT	2	1	24.7	10.0628	-34.7628	0
JOINT	2	2	-18.5742	-13.0075	-35.6502	67.232
JOINT	2	3	-9.19554	-7.35552	0	16.5511
JOINT	3	1	1.9392	0	0	0
JOINT	3	2	-10.369	0	0	0
JOINT	3	3	-6.76996	0	0	0

```
DISPLACEMENTS

JOINT ROTATIONS (RADIANS)
```

	I	J	
JOINT	1	1	9.93845E-3
JOINT	1	2	-2.92789E-3
JOINT	1	3	-1.21318E-3
JOINT	2	1	8.63133E-3
JOINT	2	2	-2.24272E-3
JOINT	2	3	-6.2215E-4
JOINT	3	1	0
JOINT	3	2	0
JOINT	3	3	0

```
HORIZONTAL DISPLACEMENTS
```

	DISPLACEMENT
ROOF LEVEL	18.9947 MM
FLOOR 1 LEVEL	11.1706 MM
FOUNDATION LEVEL	0 MM

IF FURTHER LOAD CASES FOR THIS FRAME TYPE 1 ELSE 0 ? 0

IF FURTHER FRAMES TO BE ANALYSED TYPE 1 ELSE 0 ? 0

RUNNING TIME: 6.5 SECS   I/O TIME :   23.1 SECS

4.4 Patterns of Loading and the Determination of Critical Force Actions

4.4.1 Introduction

All designers, both manual and computer aided, are interested in knowing the maximum values that force actions are likely to take during the life of a structure. And although there are exceptions in the approach of each group to the problem, broadly speaking, manual designers are concerned with devising a *limited* number of loading patterns which will reasonably approximate to the desired result for *all* the force actions, whilst programmers are probably influenced by a concern with simple programming procedures. The fact that this is also compatible with achieving an 'exact' solution is a happy coincidence. It would not be possible in a general program written for the design of irregular frames to program for the *automatic* selection of a *limited* number of useful load patterns. The only solution in this case would be to transfer the responsibility for the selection of the loading patterns to the program user. It is, however, a simple matter to automatically simulate every possible loading arrangement, regardless of its significance. Or to be more precise, it is a simple matter to examine the effect of every possible loading arrangement on a particular force action. In this way the programmer is not so much concerned with specifying significant loading patterns as determining the effect of a single loaded member on all the forces in the structure.

The effects of a system of superimposed beam loads may be investigated by analysing the structure for one member loaded at a time. Stored values of the force actions induced by individual loads may then be treated as influence coefficients from which the envelope forces are obtained on summing appropriate values.

A significant increase in computing time will inevitably accompany an exhaustive treatment of the loading system. But this aspect is better judged in relation to a particular problem. When the structural steelwork design Program SFD1 (see Chapter 6) is run on an ICL 1904S computer one complete design iteration for a 3-bay, 4-storey frame takes only 40 seconds. This calculation comprises a frame analysis for each of 13 individual loading cases, the determination of design force actions and the design of 28 members. At the expense of extra programming effort, recognition of structural symmetry where it exists would lead to a substantial reduction in computing time. This is achieved without affecting the accuracy of the solution. Further economies in computing time could result from incorporating acceptable approximations into the program; for example by recognizing that for most practical purposes the effects of a single loaded member on the bending of other members in the structure are confined to its near neighbours.

4.4.2 The Determination of Influence Coefficients and their Application in Design

A common loading system for building structures comprises the self-weight of the structure (the floors, beams, columns, partition walls, etc.) acting in conjunction with combinations of uniformly distributed beam superimposed loads. At some stage in the design wind loading is also considered and this is often assumed to be in the form of point loads applied at roof and floor levels. A complete investigation of the effects of this loading system would require as many individual analyses as there are beams carrying superimposed loads plus a further one to take account of wind effects. If it is not possible to simulate the self-weight analysis by adding together suitably factored superimposed load solutions then a formal self-weight analysis is also necessary. Assuming that the frame shown in Fig. 4.8a is loaded in this way then seven separate analyses are required to fully describe its response to loads.

Perhaps it would be more accurate to think in terms of separate loading conditions rather than separate analyses. It is only in a relaxation type of solution that the whole of the analytical procedure must be followed through for each loading condition. When analysis is based upon the formal solution of the set of equations $p = Kd$ (see Section 4.3.2), by specifying p (i.e. P( )) as a two-dimensional array then any number of load conditions may be accommodated whilst the conversion of K into the form given by the set of equations (4.7) is only carried out once.

In a multi-load condition analysis the force arrays must take a different form from those discussed in Section 4.2.1. There it was assumed that a two-dimensional array of the type FORCE$(S + 1, B + 1)$—where $(S + 1)$ and $(B + 1)$ refer to the number of rows and columns respectively—was needed to record the values of one type of force action. But if a complete analysis now requires the investigation of NLC load conditions then a total of $(S + 1)*(B + 1)*NLC$ elements must be reserved to record the values of one force action type for all load conditions. If the programming language is Algol then it is expedient to use the three-dimensional array facility offered by that language. A FORCE array would then take the form FORCE$(S + 1, B + 1, NLC)$. A notional view of such an array is shown in Fig. 4.8b. Each layer of this array conveniently holds the values of a force action type resulting from a single load condition. Each vertical column of elements therefore represents a set of influence coefficients for that force action at a specific location in the structure. The alternative is to use an expanded two-dimensional array which takes the form FORCE $(S + 1, (B + 1)*NLC)$ as shown in Fig. 4.8c. With this arrangement each block

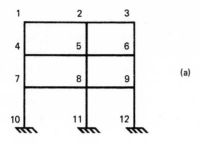

(a)

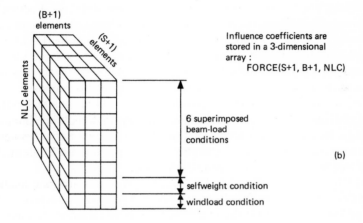

Influence coefficients are stored in a 3-dimensional array :
FORCE(S+1, B+1, NLC)

6 superimposed beam-load conditions

selfweight condition

windload condition

(b)

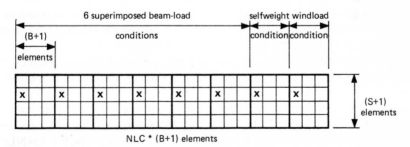

(c)

Figure 4.8 Storage of influence coefficients

of $(S + 1)*(B + 1)$ elements in turn contains the values of a force action type resulting from a single load condition. The elements at $(B + 1)$ intervals along a row of the force array now represent the influence coefficients for a particular location in the structure. In Fig. 4.8c the elements containing x might relate to the East Moment (EM) at joint number 4, i.e. the moment at the left hand end of member 4,5. The moment at this location due to superimposed loads on all spans plus the effect of self-weight is obtained by summing algebraically the contents of the first seven marked elements. The maximum hogging moment would result from summing the content of the self-weight element and all the marked elements whose contents have the same sign as this.

A block of program given at the end of this section demonstrates the way in which FORCE array information may be manipulated to determine critical force actions. The specific problem is to evaluate the maximum moment (due to superimposed beam loads and self-weight) in each column of a B-bay, S-storey frame and to store these quantities in an array M( ). For this frame the number of load conditions is $NLC = B*S + 2$. It is assumed that the program prior to line 1470 analysed the frame and stored the moments induced at the top and bottom of each column for each load condition in the force arrays S( ) and N( ). These arrays are of the form shown in Fig. 4.8c.

Each structural column, identified by its (I, J) location in the frame, is considered in turn. If we were dealing with a single load condition then the force array elements associated with a particular column would be $S(I, J)$ and $N(I + 1, J)$. But in this problem the force actions resulting from $(NLC - 2)$ load conditions must be considered. Consequently there is no direct correspondence between the location of a column in the frame and the location of its associated moments in the force arrays. This is overcome by referring to the associated elements as $S(I, Z)$ and $N(I + 1, Z)$, where Z is a function of the load condition reference number (X), the number of row elements in each sub-block $(B + 1)$ and the value of J for the column under consideration. At line 1540, Z is set equal to $(X - 1)*(B + 1) + J$, thus for the first load condition $Z = J$; for subsequent load conditions the elements are called at $(B + 1)$ intervals.

The overall procedure comprises three basic operations. The first is executed at lines 1530 to 1630 where, for the first $(NLC - 2)$ load conditions, the contents of the force array elements associated with column (I, J) are added to the contents of the variables A, B, C, or D. A records the sum of the contents of elements storing positive moments, and B the sum of the negative moments, considered in array S( ). In a similar way C and D are assigned the moments from array N( ). At lines 1640 to 1680 the second operation adds the relevant self-weight moment to A, B, C, and D. In the final operation, at lines 1690 to 1760, the contents of A, B, C, and D are compared in order to determine the largest numerical value. This represents the greatest moment to which the column will be subjected under the specified loading conditions and the value is recorded in $M(I, J)$.

In a working program we would also be interested in knowing the magnitude of the axial load which accompanied the maximum moment. It is expedient to determine both of these quantities simultaneously. If the variables A1, B1, C1, and D1 were nominated to store axial loads selected from a force array Q( ), say, then by introducing additional statements of the form:

1565 LET A1 = A1 + Q(I, Z)

each time that a moment (in this case $S(I, Z)$) was added to A, the axial load accompanying this moment would be added algebraically to the existing value of A1. Further statements of the form:

    1725 LET $F(I, J) = A1$

would record this axial load value for use in the event that A was found to contain the largest moment.

To determine the numerically largest bending moment in each column of a B-bay, S-storey frame:

```
1470 FOR I = 1 TO S
1480 FOR J = 1 TO B + 1
1490 LET A = 0
1500 LET B = 0
1510 LET C = 0
1520 LET D = 0
1530 FOR X = 1 TO NLC − 2
1540 LET Z = (X − 1)*(B + 1) + J
1550 IF S(I, Z) < 0 THEN 1580
1560 LET A = A + S(I, Z)
1570 GOTO 1590
1580 LET B = B + S(I, Z)
1590 IF N(I + 1, Z) < 0 THEN 1620
1600 LET C = C + N(I + 1, Z)
1610 GOTO 1630
1620 LET D = D + N(I + 1, Z)
1630 NEXT X
1640 LET Y = (NLC − 2)*(B + 1) + J
1650 LET A = A + S(I, Y)
1660 LET B = B + S(I, Y)
1670 LET C = C + N(I + 1, Y)
1680 LET D = D + N(I + 1, Y)
1690 IF A > ABS(B) THEN 1720
1700 LET M(I, J) = ABS(B)
1710 GOTO 1730
1720 LET M(I, J) = A
1730 IF M(I, J) > C THEN 1750
1740 LET M(I, J) = C
1750 IF M(I, J) > ABS(D) THEN 1770
1760 LET M(I, J) = ABS(D)
1770 NEXT J
1780 NEXT I
```

List of Variables and Arrays used in program extract given on page 141

A   Maximum +ve South moment
B   Maximum −ve South moment
C   Maximum +ve North moment
D   Maximum −ve North moment
I   Counter
J   Counter
M( ) Array of maximum column moments
N( ) Array of moments at bottom end
      of column
S ( ) Array of moments at top
      end of column
X   Identifies load condition
Z   Location in row of element from
      N( ) or S( )

4.5 A Computing Equivalent of Design Experience

4.5.1 Introduction

The need to save time by utilizing previous experience to anticipate the solution to a problem is not usually a pressing one in the case of computer aided design. The time taken by a computer to execute a formal design calculation from first principles is often trivial even when compared with that taken by a manual designer pursuing an approximate solution. But given the situation where there are a number of problems to solve in the same general area, possibly over a long period of time, and the computing time to produce an individual solution is a significant cost factor, then it is useful to consider what help past solutions might contribute towards accelerating the computer aided design process.

The processes of building up experience in a particular field, recalling it from memory, assessing its relevance and taking an appropriate course of action are ones which can be reproduced (insofar as they *have* a logical interpretation) by a computer system which allows a program and information held permanently on file to interact.

Both manual and computer aided operations exploit some source of design experience whenever they make reference to tabulated design coefficients. But in the present context we are concerned more with 'personal' experience, which itself might be the product of a number of those specialized sources. From a fund of personal experience the manual designer can often find a short-cut to the solution of his problem by interpolating between remembered results. An analogous computing situation is to locate a wanted quantity within an array of numerical information which itself is the product of previous program runs. If the contents of individual elements represent past solutions then, taken together, they constitute experience.

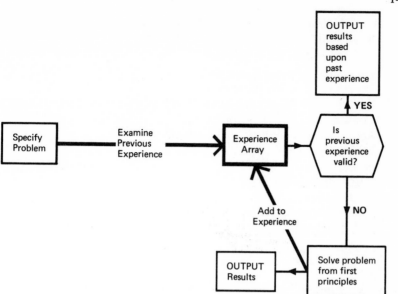

Figure 4.9 Flow diagram showing the growth and use of experience

A designer's experience is complete if the solution to any future problem lies within its scope. The computing equivalent of this state is to have access to an array in which the range of parameters is wide enough to envelop future problems and at the same time the increments between neighbouring elements are sufficiently small to justify interpolation. But in the same way that a designer may need to widen the range of his experience by occasionally solving a problem from first principles, some regions within a design experience array may also need to be expanded. The computer would recognise this situation when, having been presented with a problem, it reviewed its own experience and concluded that an interpolated solution was unacceptable. The program would then pursue a formal solution, the results of which would not only be output in the normal way but would also be transferred to a permanent file for future reference. This process is summarized in the flow diagram shown in Fig. 4.9 and is further amplified in the description of Program EX2 which follows.

4.5.2 Program Specification—EX2

This program was written to demonstrate one way in which the growth of experience may be recorded in a computer's memory and how criteria could be introduced which would determine whether that experience had matured sufficiently to be used.

In a working system the experience array would need to be part of the permanent filestore because only then would it be possible to preserve the contents of that array for future reference. Even so, the principles of the method may still be demonstrated on a less sophisticated computer system if the loss of 'experience' at the end of a program run is acceptable.

4.5.3 Description of Program EX2

A list of the variables and arrays used is given below, followed by the program listing.

A ⎫
B ⎬ Counters
C ⎭
E( ) Experience array

J Counter
K Switch
X Variable parameter
Y Wanted quantity

```
100 PRINT "THIS PROGRAMME GIVES Y FOR DIFFERENT VALUES OF"
110 PRINT "THE PARAMETER X IN THE EXPRESSION Y=X↑2-3*X+4"
120 PRINT
130 DIM E(2,100)
140 FOR J=1 TO 100
150 E(1,J)=10↑6
160 NEXT J
170 PRINT "YOUR VALUE OF X IS";
180 INPUT X
190 FOR A=1 TO 100
200 IF A>2 THEN 220
210 IF E(1,A)=10↑6 THEN 420
220 IF E(1,A)=X THEN 440
230 IF E(1,A)<X THEN 340
240 IF A=1 THEN 260
250 IF E(1,A)-E(1,A-1)<=0.5 THEN 470
260 IF E(1,A+1)=10↑6 THEN 290
270 IF E(1,A)-X>0.5 THEN 290
280 IF E(1,A+1)-E(1,A)<=0.5 THEN 490
290 IF A<3 THEN 320
300 IF X-E(1,A-1)>0.5 THEN 320
310 IF E(1,A-1)-E(1,A-2)<=0.5 THEN 510
320 IF E(1,A)=10↑6 THEN 420
330 GOTO 350
340 NEXT A
350 FOR B=A TO 100
360 IF E(1,B)=10↑6 THEN 380
370 NEXT B
380 FOR C=B TO A+1 STEP -1
390 E(1,C)=E(1,C-1)
400 E(2,C)=E(2,C-1)
410 NEXT C
420 E(1,A)=X
430 E(2,A)=X↑2-3*X+4
440 PRINT "THEREFORE Y="E(2,A)
450 PRINT
460 GOTO 540
470 Y=E(2,A-1)+(X-E(1,A-1))*(E(2,A)-E(2,A-1))/(E(1,A)-E(1,A-1))
480 GOTO 520
490 Y=E(2,A)-(E(2,A+1)-E(2,A))*(E(1,A)-X)/(E(1,A+1)-E(1,A))
500 GOTO 520
510 Y=E(2,A-1)+(E(2,A-1)-E(2,A-2))*(X-E(1,A-1))/(E(1,A-1)-E(1,A-2))
520 PRINT "THEREFORE Y="Y
530 PRINT
540 PRINT
550 PRINT "IF ANOTHER SOLUTION IS REQUIRED THEN TYPE 1 ELSE 0";
560 INPUT K
570 IF K=1 THEN 170
```

```
580 PRINT
590 PRINT "CONTENTS OF EXPERIENCE ARRAY ***************"
600 PRINT
610 PRINT "  X          Y"
620 FOR J=1 TO 100
630 IF E(1,J)=10↑6 THEN 660
640 PRINT E(1,J)"      "E(2,J)
650 NEXT J
660 END
```

4.5.3.1 Block (A): Statements 100–180

In this program the expression $Y = X^2 - 3X + 4$ has been arbitrarily chosen to stand proxy for some aspect of design which would otherwise need considerable computing time to evaluate.

The dimensions of the 'experience array' E( ) are set at line 130. It has two rows (the first stores values of X and the second, corresponding values of Y) and 100 columns. The latter dimension is assumed to be large enough to encompass the required range of experience. At lines 140 to 160 the elements in the first row of E( ) are set to 10^6, an arbitrarily high value which is outside the proposed range of experience for values of X. In the ensuing procedure a first row element content of 10^6 indicates that the corresponding second row element does not yet hold a key solution, that is one which stems from an accurate calculation.

The parameter X is input at lines 170 to 180.

4.5.3.2 Block (B): Statements 190–510

Experience is often gained in a random fashion, but before it can be utilized it must be rearranged in an orderly way; this is one purpose of Block (B). The other is to assess the current state of experience as each new problem is proposed, with the object of deciding whether a formal solution or an approximate one should be pursued. In Program EX2 these goals are met by allowing for four possibilities. They are that the proposed solution is:

1. a key solution which will increase the range of experience and be inserted into its correct slot in array E( );
2. one which is already bracketed by two existing solutions which themselves are close enough to permit linear interpolation;

or 3 and 4 one which is close enough to a pair of closely related existing solutions to allow either backward or forward linear extrapolation.

In 1 the programmed criterion which, if met, will permit an interpolated solution is that the difference between the bracketing values of X must not be greater than 0.5. In 3 and 4 the pair of neighbouring X values must be within 0.5 of one another, and in addition the new value of X must be within 0.5 of at least one of the pair. If allowable increments between values of X were specified as variables then their choice could be left to the program user. For each of the cases 2, 3 and 4, since the ensuing solutions would be approximate ones they would not be recorded in the experience array.

The way in which the information in array E( ) is arranged and assessed is most easily seen by taking numerical examples and tracing the way in which the statements between lines 200 to 330 affect the course of a solution. Let us suppose then that when a solution for $X = 2.9$ is proposed the state of the experience array is already:

1.8	2.7	3.1	10^6	10^6	etc.
1.84	3.19	4.31	0	0	etc.

$E( )$

The FOR and NEXT statements at lines 190 and 340 respectively say that each E( ) array row element will be considered in turn. Thus:

At line 190: $A = 1$ $A = 2$ $A = 3$
At line 200: $A \not> 2$ $A \not> 2$ $A > 2$
At line 210: $E(1,1) \neq 10^6$ $E(1,2) \neq 10^6$ Therefore skip line 210
At line 220: $E(1,1) \neq 2.9$ $E(1,2) \neq 2.9$ $E(1,3) \neq 2.9$
At line 230: $E(1,1) < 2.9$ $E(1,2) < 2.9$ $E(1,3) \not< 2.9$
At line 240: —   —   —   —   —   —   —$A \neq 1$
At line 250: —   —   —   —   —   —   —$E(1,3) - E(1,2) < 0.5$
    Therefore GOTO 470 and
    interpolate the solution
    between $E(2,3)$ and $E(2,2)$

Because the solution for $X = 2.9$ is interpolated the contents of E( ) will be unchanged.

If a solution is now required for $X = 2.0$ then:

At line 190: $A = 1$ $A = 2$
At line 200: $A \not> 2$ $A \not>$
At line 210: $E(1,1) \neq 10^6$ $E(1,2) \neq 10^6$
At line 220: $E(1,1) \neq 2.0$ $E(1,2) \neq 2.0$
At line 230: $E(1,1) < 2.0$ $E(1,2) \not< 2.0$
At line 240: —   —   — $A \neq 1$
At line 250: —   —   — $E(1,2) - E(1,1) > 0.5$ Therefore cannot
                             interpolate

At line 260: —   —   — $E(1,3) \neq 10^6$
At line 270: —   —   — $E(1,2) - 2.0 > 0.5$ Therefore cannot
                             back extrapolate

     JUMP
At line 290: —   —   — $A < 3$ Therefore cannot
                             forward extrapolate

JUMP
At line 320: — — — $\quad$ $E(1,2) \neq 10^6$
At line 330: — — — $\quad$ *Therefore GOTO* 350

In the above example the programmed calculation jumps to line 350 when X has a value such that existing experience is of no help in achieving a solution. The value of Y for X = 2.0 is therefore an addition to existing experience which must be inserted into E( ); but before this can be done the contents of E( ) must be rearranged in a two-stage operation. In the first stage (see lines 350 to 370) the first row of E( ) is searched to find the location of the first element with a content of 10^6—i.e. $E(1, B)$, where in this case B = 4. Since it has already been shown that the current value of X lies between $E(1, 1)$ and $E(1, 2)$ the second stage must be to make room in the array for the new solution. This is done by moving the contents of all the elements embraced by $E(1, 2)$ and $E(1, B-1)$ one element to the right, a process which is carried out in reverse order (see lines 380 to 410) so as not to overwrite existing information. After this operation array E( ) takes the form:

1.8	2.7	2.7	3.1	10^6	etc.
1.84	3.19	3.19	4.31	0	etc.

$$E(\ )$$

Following this the current values of X and Y are entered into E( ) at lines 420 and 430. The updated version of the experience array is now:

1.8	2.0	2.7	3.1	10^6	etc.
1.84	2.0	3.19	4.31	0	etc.

$$E(\ )$$

In the present state of array E( ), if a solution for X = 1.4 is sought then it will be a result of back extrapolation from the current 1st and 2nd elements. Thus:

At line 190: A = 1
At line 200: A ≯ 2
At line 210: $E(1, 1) \neq 10^6$
At line 220: $E(1, 1) \neq 1.4$
At line 230: $E(1, 1) > 1.4$
At line 240: A = 1 therefore skip line 250
At line 260: $E(1, 2) \neq 10^6$
At line 270: $E(1, 1) - X < 0.5$
At line 280: $E(1, 2) - E(1, 1) < 0.5$ $\quad$ therefore GOTO 490 and back
$\qquad\qquad\qquad\qquad\qquad\qquad\qquad$ extrapolate from $E(1, 2)$ and $E(1, 1)$

Since this is an approximate solution then the contents of array E( ) will be unaffected.

It is interesting to note that had the problem for X = 2.0 and X = 1.4 been solved in reverse order then the solution for X = 1.4 would have been an exact one, and as such would have become part of the experience array. On the other hand, the solution for X = 2.0 would have been a result of forward extrapolation from the X = 1.4 and X = 1.8 solutions, and would not therefore have added to existing experience.

4.5.3.3 Block (C): Statements 520–560

This block of statements, which completes the program, causes the results of individual problems to be printed out and allows further problems to be solved. Following the last solution of a program run the contents of the experience array are printed out; this serves as a check that the program is behaving correctly.

4.5.4 Example of Program EX2 Output

A typical set of solutions given by this program appears below. It is left to the reader to check that according to the problems solved, and their order of solution, the experience array takes its correct form.

```
EX2

THIS PROGRAMME GIVES Y FOR DIFFERENT VALUES OF
THE PARAMETER X IN THE EXPRESSION Y=X↑2-3*X+4

YOUR VALUE OF X IS ?  10.75
THEREFORE Y= 87.3125


IF ANOTHER SOLUTION IS REQUIRED THEN TYPE 1 ELSE 0 ? 1
YOUR VALUE OF X IS ? -5.4
THEREFORE Y= 49.36


IF ANOTHER SOLUTION IS REQUIRED THEN TYPE 1 ELSE 0 ? 1
YOUR VALUE OF X IS ? -5.7
THEREFORE Y= 53.59


IF ANOTHER SOLUTION IS REQUIRED THEN TYPE 1 ELSE 0 ? 1
YOUR VALUE OF X IS ? -4.9
THEREFORE Y= 42.31


IF ANOTHER SOLUTION IS REQUIRED THEN TYPE 1 ELSE 0 ? 1
YOUR VALUE OF X IS ? 10.3
THEREFORE Y= 79.19


IF ANOTHER SOLUTION IS REQUIRED THEN TYPE 1 ELSE 0 ? 1
YOUR VALUE OF X IS ? 9.9
THEREFORE Y= 71.97
```

IF ANOTHER SOLUTION IS REQUIRED THEN TYPE 1 ELSE 0 ? 1
YOUR VALUE OF X IS ? 10.5
THEREFORE Y= 82.8

IF ANOTHER SOLUTION IS REQUIRED THEN TYPE 1 ELSE 0 ? 0

CONTENTS OF EXPERIENCE ARRAY **************

X	Y
-5.7	53.59
-5.4	49.36
10.3	79.19
10.75	87.3125

RUNNING TIME: 2.0 SECS   I/O TIME :   6.5 SECS

Chapter 5

The Computer Aided Design of Continuous Reinforced Concrete Frames

5.1 Factors Affecting the Choice of a Design Method

Three factors dominate the character of programs written to design continuous reinforced concrete frames. They are a function of the nature of the structural material and its traditional usage. One factor is the high unit weight of concrete which, because of its substantial contribution towards the total structural loading, has a significant effect upon the final section dimensions. Another factor arises from the difficulty of choosing practical section dimensions from the infinity of combinations of concrete and steel which would theoretically satisfy the design force actions. And finally a problem common to all continuous structures irrespective of the material from which they are fabricated, but more often met with in concrete since continuity is a natural form—that of assessing the second moments of area at an early stage in the design process with sufficient accuracy that the assumed sections may indeed develop the design force actions to which they are subjected.

These factors are interrelated and their individual significance will vary from one problem to another. Manual designers are able to arrive at safe, economical solutions in finite time by broadly anticipating the final result from past experience of similar structures. An important consideration in programming for computer aided designs is therefore to decide upon the circumstances in which designer intervention becomes advantageous to the course of the solution.

The necessity for such intervention is virtually eliminated if a wholly automatic design approach, along the lines discussed in Chapter 6, is chosen. But this imposes a major constraint on the solution since the choice of sections would be limited to those included in a standard list. Fewer constraints are offered by an alternative automatic design approach—one in which the section proportions of the members, together with their steel ratios, are defined by the designer as parameters to the problem. This option is discussed in Section 5.2. It is an attractive one to pursue because whilst it offers the designer a measure of responsibility in dictating the form of the final result the total designer time involved in the operation is reduced to a minimum.

Less stereotyped total design solutions follow if the programmer subordinates his own role and allows the designer even greater freedom to influence the course of the solution. This suggests a decision design based program which, broadly speaking, can take one of two different forms. These are either to create a library of programs along the lines discussed in Chapter 3, or to write a single all-embracing program within which provision is made for the program user to communicate his decisions to the machine at suitable access points.

The first of these alternatives is discussed, in general terms, in Section 5.3. Briefly, the advantages of this approach lie in the flexibility of its application and the fact that individual programs in the library are self-contained and relatively short in length. Some aspects of the total design process may themselves be the subject of a designer decision before the calculation is begun if the library offers the designer a choice of more than one method of analysis or design technique. But unless the on-line computer system allows data and results from one program to be held on file for automatic retrieval and use at another stage in the design process then the library approach is at a distinct disadvantage due to the large amounts of information which need to be handled by the designer at each stage.

The program which is discussed in Section 5.4 illustrates in some detail the second decision design option. By virtue of the extra statements needed to input decisions and output provisional results, programs written along these lines will always be of a greater length than their automatic design equivalents—indeed, this is probably their main disadvantage and there is a danger that with small computers such programs will exceed the capacity of the machine. Any advantage in using an all-embracing type of decision design program will be lost if it does not offer the same flexibility of approach as an equivalent design library. Achieving this end necessitates a careful appraisal of the overall design process and its translation into programming terms.

Whilst it is evident that wholly automatic design or decision design based programs are possible, the programmer should always approach his problem with an open mind and appreciate that in practice there are levels of computer aided design where a formal division between the two becomes blurred. A trivial example of this is seen in the decision design based Program RCF25 (see Section 5.4) where the slabs are designed automatically with no further prompting from the designer than his decision concerning the minimum acceptable slab depth.

5.2 The Automatic Design of Reinforced Concrete Frames

5.2.1 Program Specification—RCF12

The prime object in writing this program was to demonstrate the feasibility of an automatic approach to the design of continuous reinforced concrete structures. With this in mind the program was treated as a pilot project and its scope was severely limited in both the type of structure it could handle and the member design criteria which it could satisfy. Designs are restricted to those for multi-bay, multi-storey frames having a regular bay width. Parameters specified by the designer dictate the basic section properties. For beams these consist of the chosen rib depth/breadth ratio and for columns, their shape (square or rectangular) and allowable steel percentage. The program generates a sequence of design iterations which ends when two consecutive solutions yield identical sets of sections. The output results comprise a listing of the final slab, rib and column section dimensions.

5.2.2 The RCF12 Flow Diagram

The flow diagram for this program is shown in Fig. 5.1. At Block (A), in addition to specifying the structural geometry, loading and characteristic stresses the designer must decide upon the basic member properties.

The roof and floor slabs are designed at Block (B). This is a one-off operation which follows the standard manual design practice of designing the slabs independently of the remainder of the main load carrying structure. Whilst the slabs provide lateral stability to the beams, and serve to transmit load to the main frames, they are assumed to have no effect on frame behaviour other than in their role as T-beam compression flanges. The total weight of the slabs is therefore accurately known at an early stage in the overall design process and this serves as a not unreasonable first approximation to the total weight of the structure.

The program makes its initial estimates of the beam and column section dimensions at Block (C). In the case of beams their overall depths are set initially to a value of Span/26 and for columns a minimum dimension of Length/60 is assumed.

Following this the calculation enters a loop which involves Blocks (D) to (I), the main body of the frame design procedure. Block (D) is the point of entry to each design iteration and within it current estimates of section dimensions serve as a basis for the calculation of second moments of area. The structure is analysed at Block (E). To simplify the analysis a single load condition is considered, that of structural self-weight acting together with the simultaneous application of imposed load on all spans. Since the symmetry of the structure and its loading preclude sway the analysis is carried out by a simple moment distribution procedure. In spite of the gross restriction placed upon loading conditions, the maximum force actions which are evaluated at Block (F) are considered to lead to, if not accurate then at least realistic, section sizes.

The beam and column sections are proportioned at Blocks (G) and (H) respectively. Beam sections are designed on the basis of 'balanced' bending conditions only, with no attention being paid to to the effects of shear and little regard given to deflection criteria. Column sections are designed according to a linear approximation of the CP 110 column design formulae.

At Block (I) the current set of section dimensions is compared with that given by the immediately previous design iteration. If the two sets are not identical then the calculation is redirected to Block (D), otherwise a solution has been reached and the program run terminates with the printing out of results at (Block (J)).

5.2.3 Description of Program RCF12

A list of the variables and arrays used in Programs RCF12 and RCF25 (for a description of RCF25 see Section 5.4) is given below. Those identified by a single asterisk appear only in Program RCF12 whilst those having a double asterisk belong to RCF25. Those common to both programs have no asterisk.

*A1 *A2 *A3 *A4	Temporary stores
A5	Moment of resistance of current column trial section
*A6	Current sum of rib and column dimensions
*A7	Previous sum of rib and column dimensions
A( )	Contains column widths
B	Number of bays
B( )	Contains beam widths
*C1	Trigger
C( )	Contains column depths
**D	Directs title heading
D( )	Contains beam depths
E( )	Contains T-beam flange widths
*F	Beam initial depth
F1	Distance between frames
F2	Characteristic concrete strength
F3	Characteristic steel strength
F( )	Contains the greatest support and span moments
G1	Number of joint balances per iteration
G( )	Contains rib depths
H( )	Contains axial column loads
I	Counter
I( )	Contains beam second moments of area
J	Counter
**J1	Minimum allowable rib depth
**J2	Minimum allowable column dimension
J( )	Contains column second moments of area
K	Counter
K( )	Contains column stiffnesses
L	Beam span
*L1	Zero shear distance
L( )	Contains column stiffnesses

M1   Slab design bending moment

M2   Slab moment of resistance

M3   Unbalanced moment at joint

*M4   Span moment

M5   Beam support moment of resistance

*M6   Beam span moment of resistance

M7   Column design moment

M8 ⎫
M9 ⎬ See Fig. 5.2

M( )   Contains beam stiffnesses

N1 ⎫
N2 ⎬ See Fig. 5.2
N3 ⎭

N( )   Contains beam stiffnesses

O( )   Contains moments at lower ends of columns

*P   Sets appropriate column in array

P( )   Contains moments at upper ends of columns—also 'same-section' groups of columns and column starting depths

*Q   Sets appropriate column in array

Q( )   Contains left hand end beam moments

R   Discriminates between external and internal columns

R1   Actual maximum column steel percentage

*R2   Rib depth/breadth ratio

**R3   Chosen maximum column steel percentage

R( )   Contains right hand end beam moments

S   Number of storeys

S1   Minimum slab depth

S2   Sum of stiffnesses at joint

S( )   Contains storey heights

T( )   Contains floor slab depths

*U   Counter

**U   Choose section input heading

*U1 ⎫
*U2 ⎬ Largest column dimensions in a group

*U3   Trigger

U( )   Contains column design information

V( )   Contains floor design loads

*W   Number of 'same-section' groups

**W   Intermediate step in reinforcement calculation

W1   Slab weight

W( )   Contains characteristic imposed loads

**X   Directs path of calculation

X1   Intermediate step in beam I calculation

*X2   Temporary store

**X6   Number of beams or columns to be modified

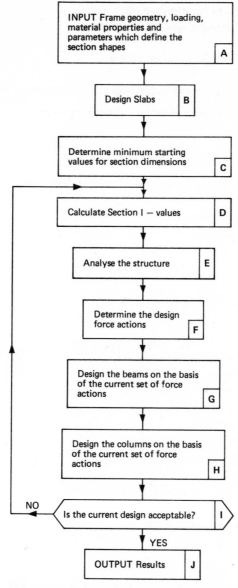

Figure 5.1 Flow diagram for program
RCF12 (the automatic design of reinforced
concrete frames)

**X7   Rib or column width
**X8   Rib or column depth
**X9   Floor level or storey number
 X( )   Contains shears at left hand end of beams
**Y   Intermediate step in reinforcement calculation—also trigger

Y( )   Contains beam design loads

**Z   Area of compression reinforcement

*Z1   Trigger

**Z1   Area of tension reinforcement

*Z2   Trigger

Z( )   Contains shears at right hand end of beams

Program RCF12 is listed below and should be read in conjunction with the flow diagram shown in Fig. 5.1.

```
100 PRINT "NUMBER OF BAYS AND STOREYS=";
110 INPUT B,S
120 PRINT "NOTE - ALL FRAME BEAM SPANS MUST BE EQUAL"
130 PRINT "SPAN (M)=";
140 INPUT L
150 PRINT "STOREY HEIGHTS (M)"
160 FOR I=1 TO S
170 IF I>1 THEN 200
180 PRINT "  TOP STOREY HEIGHT=";
190 GOTO 210
200 PRINT "HEIGHT OF STOREY"S-I+1"=";
210 INPUT S(I)
220 NEXI I
230 PRINT "DISTANCE BETWEEN FRAMES (M)=";
240 INPUT F1
250 PRINT "MINIMUM SLAB DEPTH (MM)=";
260 INPUT S1
270 PRINT "CHARACTERISTIC IMPOSED LOADS (KN/M+2)"
280 PRINT "AT ROOF AND FLOOR LEVELS"
290 FOR I=1 TO S
300 IF I>1 THEN 330
310 PRINT "       ROOF LOAD=";
320 GOTO 340
330 PRINT "LOAD AT FLOOR"S-I+1"=";
340 INPUT W(I)
350 NEXI I
360 PRINT "REQUIRED RIB DEPTH/BREADTH RATIO AT ALL LEVELS=";
370 INPUT R2
380 PRINT "IF SQUARE COLUMNS REQUIRED TYPE 1 ELSE 0";
390 INPUT C1
400 PRINT "IF YOUR DESIGN INCLUDES GROUPS OF COLUMNS HAVINC"
410 PRINT "THE SAME SECTION THEN TYPE 1 ELSE 0";
420 INPUT Z1
430 IF Z1=0 THEN 680
440 PRINT "IF INFORMATION IS REQUIRED CONCERNING A TYPICAL"
450 PRINT "RESPONSE TO SAME-SECTION DATA INPUT"
460 PRINT "THEN TYPE 1 ELSE 0";
470 INPUT Z2
480 IF Z2=0 THEN 600
490 PRINT
500 PRINT "?  1,4    NUMBERS ON THE LEFT REPRESENT THE OUTER"
510 PRINT "?  1,4    LINE OF COLUMNS - THOSE ON THE RIGHT"
520 PRINT "?  2,4    REPRESENT ALL THE INNER COLUMN LINES."
530 PRINT "?  2,5    IF B=1 THEN INPUT LEFT HAND LINE ONLY."
540 PRINT "?  2,5    ALL COLUMNS ASSIGNED THE SAME NUMBER"
550 PRINT "?  3,6    TAKE THE DIMENSIONS OF THE LARGEST"
560 PRINT "?  3,6    SECTION IN THE GROUP."
570 PRINT
580 PRINT "IN THIS EXAMPLE S=7 AND THE NUMBER"
590 PRINT "OF SAME-SECTION GROUPS=6"
600 FOR I=1 TO S
610 IF B=1 THEN 640
```

```
620 INPUT P(I,B+3),P(I,B+4)
630 GOTO 650
640 INPUT P(I,B+3)
650 NEXT I
660 PRINT "NUMBER OF SAME-SECTION GROUPS=";
670 INPUT W
680 PRINT "MAXIMUM COLUMN STEEL PERCENTAGE=";
690 INPUT R1
700 PRINT "FCU AND FY=";
710 INPUT F2,F3
720 PRINT "NUMBER OF INDIVIDUAL JOINT BALANCES/ITERATION=";
730 INPUT G1
740 FOR I=1 TO S
750 IF F1*1000/26+30> S1 THEN 780
760 T(I)=S1
770 GOTO 790
780 T(I)=F1*1000/26+30
790 W1=T(I)*24/1000
800 V(I)=1.4*W1+1.6*W(I)
810 M1=V(I)*F1+2/9
820 M2=0.15*F2*(T(I)-30)+2/1000
830 IF M2>=M1 THEN 860
840 T(I)=T(I)+10
850 GOTO 790
860 IF T(I)/10=INT(T(I)/10) THEN 880
870 T(I)=INT(T(I)/10)*10+10
880 NEXT I
890 PRINT
900 PRINT
910 FOR I=1 TO S
920 D(I)=L*1000/26+60
930 F=D(I)
940 B(I)=(D(I)-T(I))/R2
950 NEXT I
960 FOR I=1 TO S
970 FOR R=1 TO 2
980 A(I,R)=S(I)*1000/60
990 C(I,R)=A(I,R)
1000 P(I,B+2)=C(I,1)
1010 NEXT R
1020 NEXT I
1030 FOR I=1 TO S
1040 E(I)=140*L+B(I)
1050 IF E(I)<1000*F1 THEN 1070
1060 E(I)=1000*F1
1070 X1=((E(I)-B(I))*T(I)+2/2+B(I)*D(I)+2/2)
1071 X1=X1/((E(I)-B(I))*T(I)+B(I)*D(I))
1080 I(I)=(E(I)-B(I))*T(I)+3/12+B(I)*D(I)+3/12
1090 I(I)=I(I)+(E(I)-B(I))*T(I)*(X1-T(I)/2)+2
1100 I(I)=I(I)+B(I)*D(I)*(X1-D(I)/2)+2
1110 NEXT I
1120 FOR I=1 TO S
1130 FOR R=1 TO 2
1140 J(I,R)=A(I,R)*C(I,R)+3/12
1150 NEXT R
1160 NEXT I
1170 FOR I=1 TO S
1180 Y(I)=1.4*(D(I)-T(I))*B(I)*24/1000000+V(I)*F1
1190 NEXT I
1200 FOR I=1 TO S
1210 FOR J=1 TO B+1
1220 IF J=B+1 THEN 1250
1230 Q(I,J)=-Y(I)*L+2/12
1240 IF J=1 THEN 1260
1250 R(I,J)=Y(I)*L+2/12
1260 NEXT J
```

```
1270 NEXT I
1272 FOR I=1 TO S+1
1273 FOR J=1 TO B+1
1274 O(I,J)=0
1275 P(I,J)=0
1276 NEXT J
1277 NEXT I
1280 FOR I=1 TO S
1290 FOR J=1 TO B+1
1300 IF J=1 THEN 1340
1310 IF J=B+1 THEN 1340
1320 R=2
1330 GOTO 1350
1340 R=1
1350 IF I=1 THEN 1370
1360 K(I,J)=J(I-1,R)/S(I-1)
1370 L(I,J)=J(I,R)/S(I)
1380 IF J=B+1 THEN 1410
1390 M(I,J)=I(I)/L
1400 IF J=1 THEN 1420
1410 N(I,J)=I(I)/L
1420 NEXT J
1430 NEXT I
1440 FOR K=1 TO G1
1450 FOR I=1 TO S
1460 FOR J=1 TO B+1
1470 M3=O(I,J)+P(I,J)+Q(I,J)+R(I,J)
1480 S2=K(I,J)+L(I,J)+M(I,J)+N(I,J)
1490 O(I,J)=O(I,J)-M3*K(I,J)/S2
1500 P(I,J)=P(I,J)-M3*L(I,J)/S2
1510 Q(I,J)=Q(I,J)-M3*M(I,J)/S2
1520 R(I,J)=R(I,J)-M3*N(I,J)/S2
1530 IF I=1 THEN 1550
1540 P(I-1,J)=P(I-1,J)-M3*K(I,J)/(2*S2)
1550 O(I+1,J)=O(I+1,J)-M3*L(I,J)/(2*S2)
1560 IF J=B+1 THEN 1590
1570 R(I,J+1)=R(I,J+1)-M3*M(I,J)/(2*S2)
1580 IF J=1 THEN 1600
1590 Q(I,J-1)=Q(I,J-1)-M3*N(I,J)/(2*S2)
1600 NEXT J
1610 NEXT I
1620 NEXT K
1630 FOR I=1 TO S
1640 X2=0
1650 FOR J=1 TO B+1
1660 IF ABS(Q(I,J))<X2 THEN 1680
1670 X2=ABS(Q(I,J))
1680 IF ABS(R(I,J))<X2 THEN 1700
1690 X2=ABS(R(I,J))
1700 NEXT J
1710 F(I,1)=X2
1720 NEXT I
1730 FOR I=1 TO S
1740 X2=0
1750 FOR J=1 TO B
1760 L1=L/2-(R(I,J+1)+Q(I,J))/(L*Y(I))
1770 M4=Y(I)*L1+2/2+Q(I,J)
1780 IF M4<X2 THEN 1800
1790 X2=M4
1800 NEXT J
1810 F(I,2)=X2
1820 NEXT I
1830 FOR I=1 TO S
1840 FOR J=1 TO B
1850 X(I,J)=-(Q(I,J)+R(I,J+1))/L+Y(I)*L/2
1860 Z(I,J+1)=(Q(I,J)+R(I,J+1))/L+Y(I)*L/2
```

```
1870 NEXT J
1880 NEXT I
1890 FOR I=1 TO S
1900 FOR J=1 TO B+1
1910 IF J=1 THEN 1950
1920 IF J=B+1 THEN 1950
1930 R=2
1940 GOTO 1960
1950 R=1
1960 IF I>1 THEN 1990
1970 H(1,J)=X(1,J)+Z(1,J)+1.4*S(1)*A(1,R)*C(1,R)*24/1000000
1980 IF I=1 THEN 2000
1990 H(I,J)=H(I-1,J)+X(I,J)+Z(I,J)+1.4*S(I)*A(I,R)*C(I,R)*24/1000000
2000 NEXT J
2010 NEXT I
2020 IF B=1 THEN 2320
2030 FOR I=1 TO S
2040 U(I,1)=0
2050 FOR J=2 TO B
2060 IF H(I,J)<U(I,1) THEN 2120
2070 U(I,1)=H(I,J)
2080 IF ABS(P(I,J))>ABS(O(I+1,J)) THEN 2110
2090 U(I,2)=ABS(O(I+1,J))
2100 GOTO 2120
2110 U(I,2)=ABS(P(I,J))
2120 NEXT J
2130 NEXT I
2140 FOR I=1 TO S
2150 A1=0
2160 A2=0
2170 FOR J=2 TO B
2180 IF ABS(P(I,J))<A1 THEN 2210
2190 A1=ABS(P(I,J))
2200 A3=J
2210 IF ABS(O(I+1,J))<A2 THEN 2240
2220 A2=ABS(O(I+1,J))
2230 A4=J
2240 NEXT J
2250 IF A1>A2 THEN 2290
2260 U(I,3)=A2
2270 U(I,4)=H(I,A4)
2280 GOTO 2310
2290 U(I,3)=A1
2300 U(I,4)=H(I,A3)
2310 NEXT I
2320 FOR I=1 TO S
2330 IF ABS(P(I,1))<ABS(O(I+1,1)) THEN 2360
2340 U(I,6)=ABS(P(I,1))
2350 GOTO 2370
2360 U(I,6)=ABS(O(I+1,1))
2370 U(I,5)=H(I,1)
2380 NEXT I
2390 FOR I=1 TO S
2400 D(I)=F
2410 M5=0.15*F2*B(I)*0.81*D(I)+2/1000000
2420 IF F(I,1)<=M5 THEN 2460
2430 D(I)=D(I)+20
2440 B(I)=(D(I)-T(I))/R2
2450 GOTO 2410
2460 M6=0.4*F2*E(I)*T(I)*(0.9*D(I)-T(I)/2)/1000000
2470 IF F(I,2)<=M6 THEN 2540
2480 D(I)=D(I)+20
2490 B(I)=(D(I)-T(I))/R2
2500 E(I)=120*L+B(I)
2510 IF E(I)<1000*F1 THEN 2530
2520 E(I)=1000*F1
```

```
2530 GOIO 2460
2540 G(I)=D(I)-T(I)
2550 IF G(I)/20=INT(G(I)/20) THEN 2570
2560 G(I)=INT(G(I)/20)*20+20
2570 IF B(I)/20=INT(B(I)/20) THEN 2590
2580 B(I)=INT(B(I)/20)*20+20
2590 NEXT I
2600 IF B=1 THEN 2720
2610 FOR K=1 TO 3
2620 IF K=3 THEN 2720
2630 IF K=2 THEN 2680
2640 P=3
2650 Q=4
2660 R=2
2670 GOIO 2750
2680 P=2
2690 Q=1
2700 R=2
2710 GOIO 2750
2720 P=6
2730 Q=5
2740 R=1
2750 FOR I=1 TO S
2760 IF K=2 THEN 2780
2770 C(I,R)=P(I,B+2)
2780 IF C1=1 THEN 2810
2790 A(I,R)=B(I)+60
2800 GOIO 2820
2810 A(I,R)=C(I,R)
2820 N3=U(I,Q)*1000/(F2*A(I,R)*C(I,R))
2830 N1=0.4+0.72*R1*F3/(100*F2)
2840 N2=0.18-0.00075*R1*F3/F2
2850 M8=0.063+0.00318*R1*F3/F2
2860 M9=0.003555*R1*F3/F2
2870 M7=U(I,Q)*1000*(0.75*S(I)*1000)+2/(1750*C(I,R))
2880 M7=M7*(1-0.0035*0.75*S(I)*1000/C(I,R))
2890 M7=(M7+U(I,P)*1000000)/(F2*A(I,R)*C(I,R)+2)
2900 IF N3>N1 THEN 2970
2910 IF N3>N2 THEN 2950
2920 A5=(M8-M9)*N3/N2+M9
2930 IF M7>A5 THEN 2970
2940 GOIO 3000
2950 A5=(N1-N3)*M8/(N1-N2)
2960 IF M7<A5 THEN 3000
2970 C(I,R)=C(I,R)+20
2980 IF C1=1 THEN 2810
2990 GOIO 2820
3000 NEXT I
3010 IF B=1 THEN 3030
3020 NEXT K
3030 FOR I=1 IO S
3040 FOR R=1 TO 2
3050 IF C(I,R)/20=INT(C(I,R)/20)THEN 3090
3060 C(I,R)=INT(C(I,R)/20)*20+20
3070 IF C1=0 THEN 3090
3080 A(I,R)=C(I,R)
3090 NEXT R
3100 NEXT I
3105 IF Z1=0 THEN 3390
3110 FOR U=1 TO W
3120 U1=0
3130 U2=0
3140 U3=0
3150 FOR I=1 TO S
3160 IF B>1 THEN 3190
3170 J=B+3
3180 GOIO 3230
```

```
3190 FOR J=B+3 TO B+4
3200 IF J=B+3 THEN 3230
3210 R=2
3220 GOTO 3240
3230 R=1
3240 IF P(I,J)><U THEN 3320
3250 IF U3=1 THEN 3300
3260 IF A(I,R)*C(I,R)<U1*U2 THEN 3320
3270 U1=C(I,R)
3280 U2=A(I,R)
3290 IF U3=0 THEN 3320
3300 C(I,R)=U1
3310 A(I,R)=U2
3320 IF B=1 THEN 3340
3330 NEXT J
3340 NEXT I
3350 IF U3=1 THEN 3380
3360 U3=1
3370 GOTO 3150
3380 NEXT U
3390 A6=0
3400 FOR I=1 TO S
3410 A6=A6+G(I)+B(I)
3420 NEXT I
3430 FOR I=1 TO S
3440 FOR R=1 TO 2
3450 A6=A6+A(I,R)+C(I,R)
3460 NEXT R
3470 NEXT I
3480 PRINT A6
3490 IF A6=A7 THEN 3520
3500 A7=A6
3510 GOTO 1030
3520 PRINT
3530 FOR I=1 TO S
3540 IF I>1 THEN 3570
3550 PRINT "        ROOF SLAB DEPTH="T(I)"MM"
3560 GOTO 3580
3570 PRINT "SLAB DEPTH AT FLOOR"S-I+1"="T(I)"MM"
3580 NEXT I
3590 PRINT
3600 PRINT "RIB DIMENSIONS"
3610 PRINT
3620 FOR I=1 TO S
3630 IF I>1 THEN 3660
3640 PRINT "   ROOF LEVEL RIB : D="G(I)"MM   B="B(I)"MM"
3650 GOTO 3670
3660 PRINT "FLOOR"S-I+1"LEVEL RIB : D="G(I)"MM   B="B(I)"MM"
3670 NEXT I
3680 PRINT
3690 FOR R=1 TO 2
3700 IF R=1 THEN 3750
3710 PRINT
3720 PRINT "INTERNAL COLUMN DIMENSIONS"
3730 PRINT
3740 GOTO 3770
3750 PRINT "EXTERNAL COLUMN DIMENSIONS"
3760 PRINT
3770 FOR I=1 TO S
3780 IF I>1 THEN 3810
3790 PRINT "TOP STOREY COLUMNS : D="C(I,R)"MM   B="A(I,R)"MM"
3800 GOTO 3820
3810 PRINT "  STOREY"S-I+1"COLUMNS : D="C(I,R)"MM   B="A(I,R)"MM"
3820 NEXT I
3830 IF B=1 THEN 3850
3840 NEXT R
3850 END
```

5.2.3.1 Block (A): Statements 100–730

The sequence of PRINT and INPUT statements comprising this block are largely self explanatory. When it is not immediately obvious to the program user how he should present information to the machine, then the rules governing this should either be specified in the program documentation or given by the program itself in the form of output (see lines 440 to 590 for an example of the latter method). The program user is given the opportunity at line 480 of by-passing this substantial amount of printed information if he is already familiar with the input requirements.

5.2.3.2 Block (B): Statements 740–900

The roof and floor slabs are assumed to span continuously between the main structural frames. This block of statements deals with their design in a summary fashion—compare these fifteen statements with the slab design Program RC4 which was discussed in Chapter 2. The design calculation is embraced by FOR and NEXT statements (see lines 740 and 880) which cause each slab, from roof to 1st floor level, to be designed in turn.

The calculation begins (at lines 750 to 780) by choosing the largest of two minimum values—either the minimum overall depth needed to satisfy the deflection criterion (i.e. (Span $*$ 1000/26 + 30)mm) or the minimum construction depth (S1). This value is recorded in the appropriate element of array T( ). An assessment is made at line 810 of the maximum moment (M1) in the continuous slab. This is followed at line 820 by a calculation of the moment of resistance (M2) of the proposed slab based upon 'balanced' conditions and an effective depth of 30 mm less than the overall depth. If at line 830 M1 is found to be greater than M2 then the slab depth is increased by 10 mm at line 840 and a jump to line 790 recycles the procedure for the modified slab depth. Otherwise, if M2 is greater than M1, the proposed slab depth is acceptable (at least according to the programmed criteria) and it is rounded up to the nearest 10 mm and recorded in array T( ) at line 870.

5.2.3.3 Block (C): Statements 910–1020

The purpose of this block of statements is to assign starting values to the section dimensions. The initial beam section sizes are allocated at lines 910 to 950. At line 920 the minimum overall beam depth is taken as Span/26 plus an allowance for d_2. Since all the spans are equal this value is recorded in a single variable (F) for repeated use later in the design calculation. The initial rib width, calculated at line 940, is a function of the designer specified rib depth/breadth ratio (R2). Because the rib at a given floor level is assumed to be of uniform section throughout its entire length the one-dimensional arrays B( ) and D( ) are sufficient to record the beam dimensions.

The initial column section dimensions are calculated and assigned at lines 960 to 1020. Whether the columns are to be of square or rectangular cross-section their initial depths and breadths are based upon Column Length/60. Two basic sets of columns are considered—external and internal. In this program the internal columns at a given floor level are each assigned the dimensions of the most heavily loaded member in that group. The external columns are considered separately. This is why arrays A( ) and C( ), which record the column depths and breadths respectively, are two-dimensional. The initial column depth values are also stored in the (B + 2)th column of array P( ), at line 1000, so that they may be referred to later in the design calculation. This shows up a minor weakness of BASIC programming language in that the number of unique array descriptions is limited to 26. When an array was needed to record the minimum column depth values it was found that the available array descriptions had all been used. Column (B + 2) of array P( ) was therefore arbitrarily set aside for this purpose.

5.2.3.4 Block (D): Statements 1030–1160

This block is the entry point to the frame design iterative procedure. Its function is to calculate the beam and column second moments of area, and for this it uses the latest estimate of the section dimensions.

The beams are processed at lines 1030 to 1110. For the purpose of estimating member stiffness they are assumed to have a uniform T-section throughout their spans. Their effective flange widths are taken to be the lesser of the values calculated at lines 1040 and 1060. The beam second moments of area are calculated in four steps and stored in array I( ). The column second moments of area, calculated at lines 1120 to 1160, are stored in array J( ).

5.2.3.5 Block (E): Statements 1170–1620

This block of program is concerned with the analysis of the structure, a procedure which can conveniently be considered in four stages. These are to calculate the beam design loads, fixed end moments and member stiffnesses for use in the final stage, a simple moment distribution process.

At each floor level the uniformly distributed beam design load is a function of appropriate rib and slab self-weights and the imposed load. The (slab + imposed load) component was calculated at line 800 and stored in V(I). At line 1180 this is now added to the current estimate of rib weight and stored in array Y( ).

Bending moments induced at the ends of members are stored in the arrays O( ), P( ), Q( ) and R( ). These relate to the North, South, East and West moments which act at the ends of members framing into four-member joints (see Fig. 4.1e). Each array has (S + 1) rows and (B + 1) columns, i.e. as many elements as there are joints in the structure. And to preserve the one-to-one relationship between the elements of these arrays some elements will necessarily have permanently zero values. For example, all the elements in the first column and the last row of array R( ) will be zero since there are no members framing into the 'West' side of the boundary joints on the LHS of the structure or at foundation level. The beam fixed end moments are calculated at lines 1200 to 1270. Conditional jumps at lines 1220 and 1240 prevent values from being entered into R( ) if J = 1, and into Q( ) if J = B + 1. Thus account is taken of the absence of members beyond the vertical boundaries of the frame.

The member stiffnesses are calculated at lines 1280 to 1430. They are stored in the arrays K( ), L( ), M( ) and N( ) which are again related to the distribution of joints in the frame. It ought to be noted that this procedure requires excessive storage space—i.e. $4*(S + 1)*(B + 1)$ elements as compared with the minimum requirement of $S*(B + 1)$ elements had the beam and column stiffnesses been considered in their two separate groups. But as in this case, it is occasionally expedient to sacrifice programming elegance in the interests of simplicity at a later stage.

The formal moment distribution procedure takes place at lines 1440 to 1620. In a manual solution it is usual to encourage convergence by operating upon the largest numerical value of the unbalanced moment each time that a joint balance is to be made. When programming the procedure for solution by computer it is simpler to consider each joint in its turn without giving consideration to the magnitude of the unbalanced moment. The programmed criterion for terminating moment distribution is that the cycle of joint balancing is executed G1 times over the whole frame (where G1 is specified by the designer and input at Block (A)). Alternatively the computing time could be minimised by arranging for the relaxation procedure to halt when the largest carry-over moment had reduced to an acceptably small value.

The joint balancing cycle is governed by the FOR statements at lines 1440, 1450, and 1460 and the NEXT statements at lines 1600, 1610 and 1620. The algebraic sum of the moments acting at the ends of members framing into a joint (which is specified by its (I, J) location in the frame) is the unbalanced moment, called M3. This is calculated at line 1470. The sum of the stiffnesses of the members meeting at that joint, called S2, follows at line 1480. The statements at lines 1490 to 1520 balance the moments at the ends of the connected members. One half of each balancing moment is carried over to the remote end of the relevant connected member at lines 1540, 1550, 1570 and 1590. In this respect the conditional statements at lines 1530, 1560 and 1580 prevent carry-over moments from being assigned to non-existent members at boundary joints.

5.2.3.6 Block (F): Statements 1630–2380

The function of this block is to determine the force actions which govern the design of an individual member or group of members. The force action criteria for a continuous beam are the maximum values of support and span moment which occur along its length. For an external column the criterion is its axial load acting in conjunction with the largest bending moment in the member. But for a group of internal columns at any one level the criterion may either be the largest axial load acting together with its corresponding maximum moment or the largest moment in the group and its corresponding axial load.

The restrictions which were imposed upon the structural form and loading automatically confine the beam design force criteria to the first internal support and the outer span. And for internal columns the two design force criteria occur simultaneously at the outer columns in the group. However, in order to illustrate the way in which maximum force actions may be found, this block of program has been expanded beyond that of immediate practical necessity.

The maximum beam support design moments are determined at lines 1630 to 1720. The continuous beams are considered in order from roof to 1st floor level by giving the variable I a value between 1 and S at line 1750. For a chosen beam the operations at lines 1650 to 1700 inspect its support moments ($Q(I, J)$ and $R(I, J)$) and at each step record the greatest absolute value of bending moment found so far, in X2. When all the support moments at one level have been inspected then the value remaining in X2 is copied into the appropriate element of array F( ) at line 1710.

The operations carried out between lines 1730 and 1820 determine the maximum span moment at each floor level. There are many similarities between the overall structure of this procedure and the one discussed in the previous paragraph, but in this instance it is the spans which are primarily referred to at each floor level (see the limits that are set on J at line 1750) rather than the supports. At line 1760 L1 gives the distance from the left hand support to the section at which the shear is zero in the chosen span. Following this at line 1770 the maximum moment occurring in that span (M4) is calculated in terms of the beam design load, the distance to zero shear and the moment at the left hand support. As the spans at a given level are considered in their turn the variable X2 is used to record the maximum value of M4 so far established (see lines 1780 and 1790). When all the spans of a continuous beam have been considered then the content of X2 is transferred to $F(I, 2)$ at line 1810. Thus the first and second columns of array F( ) hold the maximum support moment and span moment design criteria respectively.

The statements at lines 1830 to 1880 comprise the first stage in determining the column design forces; this is the calculation of the beam end-shears. For each span in the frame the end-shears (which are calculated by using the span, load, and end-moments) are entered into the arrays X( ) and Z( ).

The axial column loads are calculated at lines 1890 to 2010. A column load is taken as being the sum of the axial load in the column immediately above the one considered, the end-shears from beams framing into the top of the column and the self-weight of the column itself. The statements at lines 1970 and 1990 calculate the loads on this basis for the top storey and intermediate storey columns respectively. The results are stored in a two-dimensional array called H( ) which has as many elements as there are column members.

The external and internal columns are designed separately and it is therefore unlikely that at a given storey level their dimensions will be identical. This is why the arrays A( ) and C( ) are two-dimensional, external column dimensions being held in the first column of each array and internal column dimensions in the second. The computer 'knows' that when $J = 1$ or $(B + 1)$ it is dealing with an external column calculation. This sets a variable called R equal to 1 at line 1950; otherwise at line 1930 $R = 2$. In this way the relevant column dimensions $A(I, R)$ and $C(I, R)$ are identified so that the correct column self-weight component of the axial load may be calculated at lines 1970 and 1990.

The force actions which define the (maximum axial load – corresponding moment) design criterion for groups of internal columns are held in array U( ), the maximum axial loads being recorded in the first column of this array and their corresponding moments in the second column. These force actions are determined at lines 2030 to 2130 in the following way. For a given storey level (i.e. group of internal columns) defined by the value of I the array element $U(I, 1)$ is set to zero at line 2040. The FOR statement at line 2050 implies that only internal columns are to be considered since J is restricted to values between 2 and B. A search through row I of array H( ), which is initiated by the FOR and NEXT statements at lines 2050 and 2110, locates the largest axial load in the group and assigns its value to $U(I, 1)$. The greatest moment corresponding to that value of axial load in the column is stored in the element $U(I, 2)$.

The force actions which govern the second design criterion for internal columns (i.e. maximum moment – corresponding axial load) are determined at lines 2140 to 2310. At each storey level (defined by I) the maximum moment occurring at the top of an internal column in that group is recorded in the variable A1 at line 2190. Its J location is assigned to A3 at line 2200. The maximum moment occurring at the lower end of an internal column, and its J location, are assigned to the variables A2 and A4 at lines 2220 and 2230 respectively. At lines 2250 to 2300 the contents of A1 and A2 are compared and the greatest of these is recorded in $U(I, 3)$. At lines 2270 or 2300 $U(I, 4)$ takes the value of either $H(I, A3)$ or $H(I, A4)$, the axial load which corresponds to the chosen maximum moment.

For a single-bay frame ($B = 1$) the conditional jump at line 2020 bypasses that part of the program concerned with the internal column force actions. The statements at lines 2320 to 2380 determine the maximum moment in each external column and record its value in $U(I, 6)$; the corresponding axial load is assigned to $U(I, 5)$.

At this point in the program the beam design moments are held in array F( ) and the column design moments and axial loads in array U( ). Because these arrays are referred to repeatedly in Blocks (G) and (H) it is useful to summarize their contents at this stage.

Array F( ), S-rows and 2-columns:

$J = 1$ Maximum beam support moments
$J = 2$ Maximum beam span moments

Array U( ), S-rows and 6-columns:

J = 1   Internal columns— Maximum axial load
J = 2   Internal columns— Largest moment corresponding to the maximum axial load
J = 3   Internal columns— Maximum moment
J = 4   Internal columns— Axial load corresponding to the maximum moment
J = 5   External columns—Axial load
J = 6   External columns—Corresponding largest moment

5.2.3.7 Block (G): Statements 2390–2590

The purpose of this block of statements is to design the beam sections on the basis of the current set of design force actions. For each continuous beam (see lines 2390 and 2590) the design is carried out in two stages. The first stage (see lines 2400 to 2450) is concerned with proportioning the support section. The second stage (see lines 2460 to 2580) checks the span section implied by this rib size and, if necessary, increases its dimensions.

At line 2400 D(I), the overall depth of the beam under consideration, is set to the value of F, the minimum allowable depth derived from deflection requirements (see line 930). This depth together with B(I) (see line 940) define the overall dimensions of the first trial section. Its moment of resistance (M5) is calculated on the assumption of 'balanced' bending conditions and an effective depth of (0.9 × overall beam depth) at line 2410. If at line 2420 F(I, 1), the support design moment, is found to be less or equal to M5 then a jump to line 2460 shows that the section is acceptable; otherwise the section depth is increased by 20 mm and the rib width is reassessed (see lines 2430 and 2440). The calculation then returns to line 2410 and the procedure is repeated. This cycle continues until the section moment of resistance exceeds the applied moment.

The support section depth derived in this way serves as a starting point for the design of the span section. M6, the span section moment of resistance, is calculated at line 2460 and compared at line 2470 with the applied moment F(I, 2). If F(I, 2) is greater than M6 then the section depth is increased by 20 mm at line 2480. This also requires a reassessment of the rib and flange widths at lines 2490 to 2520. With a return to line 2460 the procedure is recycled, a process which continues until the span section moment of resistance is greater than the applied moment. The rib depth and breadth are then rounded up to the nearest 20 mm at lines 2540 to 2580.

It should be noted that only during the final overall frame design iteration do the design force actions for all the members exactly match those for which the sections ought to have been proportioned. The design forces, influenced as they are by the structural self-weight, only accurately reflect that self-weight when the structural sections required by the design criteria are the same as those on which the design loads, and hence the design forces, were based. The first time that this can happen is during the second of two consecutive design iterations which both produce identical section sizes.

5.2.3.8 Block (H): Statements 2600–3380

In a multi-bay frame the internal columns are designed to resist two combinations of axial load and bending moment; the external columns are designed to resist only one combination. These six force actions and their locations in array U( ) were discussed in Section 5.2.3.6. A variable called K, set in turn to a value of 1, 2, or 3, determines at any one time which of the three sets of force actions is to be considered. Thus when K equals 1, P and Q are set to 3 and 4 respectively (see lines 2640 and 2650). This indicates that an internal column will initially be proportioned to withstand the maximum internal column moment acting together with its corresponding axial load. In addition to this, for K equal to 1 or 2, the variable R is set equal to 2 (see lines 2660 and 2700); and when K equals 3, then R is set equal to 1 (see line 2740). This allows the program to discriminate between internal and external column dimensions whilst still referring generally to these dimensions as $C(I, R)$ and $A(I, R)$.

When K is equal to 1, all the internal columns are designed to meet one set of requirements at lines 2750 to 3000. When K is equal to 2, with a single exception, a similar procedure is followed for the second force action combination. The exception is that the statement at line 2760 bypasses the column depth starting value given by $P(I, B + 2)$ (see line 1000) and chooses instead the current calculated value, $C(I, R)$. This ensures that the greatest dimensions resulting from either set of force actions are the ones which are finally recorded.

When K is equal to 3 the external columns are only once subjected to the procedure between lines 2750 and 3000 since they are only required to be designed to meet a single force action combination.

A switch called C1, which is set by the designer at the parameter input stage to 1 if the columns are to be square, or to 0 if they are to be rectangular, directs the assessment of column width to lines 2790 or 2810 respectively. For a rectangular column its width is made 60 mm greater than the rib width into which it frames. For square columns the width is made equal to the depth.

The strength of a concrete column may be checked by determining the position of a point defined by the axial design load P and the design bending moment M relative to the axial load–moment interaction envelope diagram for the member. A point lying within the area which is defined by the envelope and its axes indicates that the column has a satisfactory reserve of strength. A simplified form of the column design envelope is shown in Fig. 5.2 where, in dimensionless terms, the boundaries are defined by linear approximations to the CP 110 column design formulae. The programmed equivalent of the force interaction envelope is defined by the coordinates $(0, N1)$, $(M8, N2)$ and $(M9, 0)$. The non-zero coordinates N1, N2, M8 and M9 are calculated at lines 2830 to 2860. The value of $P/f_{cu}bd$ for the proposed column (i.e. N3) is calculated at line 2820. If at line 2900 N3 is found to be greater than N1 then the column depth must be too small, consequently a jump to line 2970 increases the depth by 20 mm. Depending upon the shape of the column cross-section a return to either of the lines 2810 or 2820 allows further envelope values to be calculated which define the strength of the new column section.

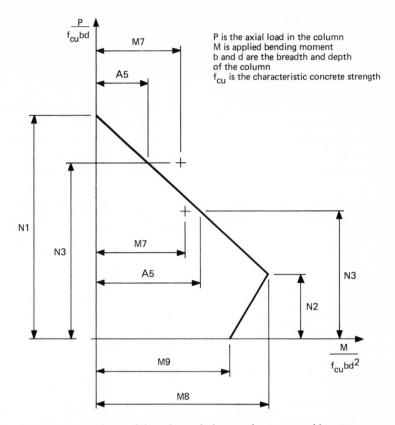

Figure 5.2 The shape of the column design envelope assumed in programs
RCF 12 and RCF 25

To take account of column slenderness the design moment is increased in accordance with CP 110 recommendations. The content of a variable called M7, which represents the modified moment divided by $f_{cu} bd^2$, is calculated at lines 2870 to 2890.

When it is established that N3 is less than N1 then the statement at line 2910 further determines whether N3 is greater or less than N2 (see Fig. 5.2). This is necessary because the assumed envelope is a non-continuous function. It then becomes possible to calculate A5, the maximum allowable value of $M/f_{cu} bd^2$ corresponding to the current value of N3 (see lines 2920 and 2950). A comparison of the values of M7 and A5 at line 2930 or 2960 then determines whether the current depth should be increased and the design cycle repeated. If not then the dimensions currently held in C(I, R) and A(I, R) are accepted and operations begin on the next column.

The design of each column section is completed with the rounding up of dimensions to the nearest 20 mm at lines 3030 to 3100.

The value of Z1, which was input at line 420, determines whether the program area at lines 3110 to 3180 is entered. This is a procedure which forces specified groups of columns to take the dimensions of the largest section in the group. These groups of same-section members were defined in Block (A) at lines 600 to 670 where it is seen that the group descriptions are stored in the elements of columns $(B + 3)$ and $(B + 4)$ of array $P(\ )$. External columns are referred to when J is equal to $(B + 3)$ and all the internal columns at one level when J is equal to $(B + 4)$.

Possible 'same-section'	1	2	1	1	7	3
array P( ) contents for	1	2	1	1	6	1
4-storey frame.	3	4	1	1	2	8
	3	4	1	1	5	4
	(a)		(b)		(c)	

Whilst groups of same-section columns may be entered into array $P(\ )$ in any order the arrangement shown at (a) above is the simplest style. The arrangement at (b) shows a single group which would force all of the columns in the structure to take the same section dimensions. The unlikely (and unnecessary) group numbering shown at (c) would allow all the columns to retain their own designed dimensions—the result which would have been achieved anyway by setting Z1 to zero at line 420.

The number of same-section groups is recorded in a variable called W (see line 670); the statement at line 3110 indicates that the 'sorting and assigning' operation will be carried out W times. The variables U1, U2, and U3 are reset to zero at lines 3120 to 3140 for each new value of U. U1 and U2 subsequently hold the dimension of the largest section in the group which is defined by the value of U. The value of U3, which later becomes 1 at line 3360, determines the order in which the sorting and assigning procedures are carried out.

For each value of U the relevant columns of array $P(\ )$ are searched twice. The object of the first search, when U3 is equal to zero, is to determine the dimensions of the largest column in the group defined by the value of U. This is done by examining the dimensions of each column in the group and successively replacing the contents of U1 and U2 with new dimensions as and when a larger column is found; see in particular the lines 3240 and 3260 to 3290. For the same value of U, but with U3 now equal to 1, a second search assigns to all the sections in the group the dimensions held in U1 and U2. The relevant statements for this procedure are at lines 3240, 3250, 3300 and 3310.

5.2.3.9 Block (I): Statements 3390–3510

This block sets the criterion by which the design cycle is terminated; a criterion which is based upon a comparison between the section dimensions produced by successive iterations. The sum of the current design rib and column dimensions is recorded in the variable A6 (see lines 3390 to 3470). The content of A6, printed out at line 3480, gives the designer some indication of the course that the solution is taking. After the first design iteration the content of A6 is recorded in A7 at line 3500. Thus at the end of subsequent iterations a comparison between the contents of A6 and A7 (see line 3490) shows whether two successive identical results have been achieved. If so then a jump to line 3520 provides an entry into Block (J); otherwise the current set of section dimensions forms the basis on which a further design iteration begins at Block (D).

5.2.3.10 Block (J): Statements 3520–3850

The information which is printed out at this block consists of a simple summary of the slab, rib and column dimensions which result from the input design parameters and the assumptions built into the programmed calculation.

5.2.4 Example of Program RCF12 Output

A typical solution given by this program appears below. The chosen example is that of a 4-bay, 8-storey frame in which the designer has specified rib depth/breadth ratios of 2.0 and columns of square cross-section with a maximum of 2% reinforcement. A further constraint on the solution is that of 'same-section' column groups. For this example a solution was achieved after 4 iterations.

```
RCF12

NUMBER OF BAYS AND STOREYS= ?  4,8
NOTE - ALL FRAME BEAM SPANS MUST BE EQUAL
SPAN (M)= ?  5.7
STOREY HEIGHTS (M)
   TOP STOREY HEIGHT= ?  3.7
HEIGHT OF STOREY 7 = ?  3.7
HEIGHT OF STOREY 6 = ?  3.7
HEIGHT OF STOREY 5 = ?  3.7
HEIGHT OF STOREY 4 = ?  3.7
HEIGHT OF STOREY 3 = ?  3.7
HEIGHT OF STOREY 2 = ?  3.7
HEIGHT OF STOREY 1 = ?  4.6
DISTANCE BETWEEN FRAMES (M)= ?  4
MINIMUM SLAB DEPTH (MM)= ?  100
CHARACTERISTIC IMPOSED LOADS (KN/M†2)
AT ROOF AND FLOOR LEVELS
        ROOF LOAD= ?  1.25
LOAD AT FLOOR 7 = ?  2.2
LOAD AT FLOOR 6 = ?  2.2
LOAD AT FLOOR 5 = ?  2.2
LOAD AT FLOOR 4 = ?  2.2
LOAD AT FLOOR 3 = ?  2.2
LOAD AT FLOOR 2 = ?  2.2
LOAD AT FLOOR 1 = ?  2.2
REQUIRED RIB DEPTH/BREADTH RATIO AT ALL LEVELS= ?  2
IF SQUARE COLUMNS REQUIRED TYPE 1 ELSE 0 ?  1
IF YOUR DESIGN INCLUDES GROUPS OF COLUMNS HAVING
THE SAME SECTION THEN TYPE 1 ELSE 0 ?  1
IF INFORMATION IS REQUIRED CONCERNING A TYPICAL
RESPONSE TO SAME-SECTION DATA INPUT
THEN TYPE 1 ELSE 0 ?  0
 ?  1,2
 ?  1,2
```

```
? 3,4
? 3,4
? 5,6
? 5,6
? 7,8
? 7,8
NUMBER OF SAME-SECTION GROUPS= ? 8
MAXIMUM COLUMN STEEL PERCENTAGE= ? 2
FCU AND FY= ? 25,250
NUMBER OF INDIVIDUAL JOINT BALANCES/ITERATION= ? 10

 13380
 14060
 14140
 14140


        ROOF SLAB DEPTH= 190 MM
SLAB DEPTH AT FLOOR 7 = 190 MM
SLAB DEPTH AT FLOOR 6 = 190 MM
SLAB DEPTH AT FLOOR 5 = 190 MM
SLAB DEPTH AT FLOOR 4 = 190 MM
SLAB DEPTH AT FLOOR 3 = 190 MM
SLAB DEPTH AT FLOOR 2 = 190 MM
SLAB DEPTH AT FLOOR 1 = 190 MM

RIB DIMENSIONS

   ROOF LEVEL RIB : D= 340 MM  B= 180 MM
FLOOR 7 LEVEL RIB : D= 360 MM  B= 180 MM
FLOOR 6 LEVEL RIB : D= 360 MM  B= 180 MM
FLOOR 5 LEVEL RIB : D= 360 MM  B= 180 MM
FLOOR 4 LEVEL RIB : D= 360 MM  B= 180 MM
FLOOR 3 LEVEL RIB : D= 360 MM  B= 180 MM
FLOOR 2 LEVEL RIB : D= 360 MM  B= 180 MM
FLOOR 1 LEVEL RIB : D= 360 MM  B= 180 MM

EXTERNAL COLUMN DIMENSIONS

TOP STOREY COLUMNS : D= 180 MM  B= 180 MM
  STOREY 7 COLUMNS : D= 180 MM  B= 180 MM
  STOREY 6 COLUMNS : D= 240 MM  B= 240 MM
  STOREY 5 COLUMNS : D= 240 MM  B= 240 MM
  STOREY 4 COLUMNS : D= 280 MM  B= 280 MM
  STOREY 3 COLUMNS : D= 280 MM  B= 280 MM
  STOREY 2 COLUMNS : D= 320 MM  B= 320 MM
  STOREY 1 COLUMNS : D= 320 MM  B= 320 MM

INTERNAL COLUMN DIMENSIONS

TOP STOREY COLUMNS : D= 260 MM  B= 260 MM
  STOREY 7 COLUMNS : D= 260 MM  B= 260 MM
  STOREY 6 COLUMNS : D= 340 MM  B= 340 MM
  STOREY 5 COLUMNS : D= 340 MM  B= 340 MM
  STOREY 4 COLUMNS : D= 400 MM  B= 400 MM
  STOREY 3 COLUMNS : D= 400 MM  B= 400 MM
  STOREY 2 COLUMNS : D= 440 MM  B= 440 MM
  STOREY 1 COLUMNS : D= 440 MM  B= 440 MM


RUNNING TIME:   36.9 SECS   I/O TIME :   40.4 SECS
```

5.3 The Program Library—A Decision Design Approach to Reinforced Concrete Frames

5.3.1 Introduction

A program library to aid the design of conventional beam and slab multi-bay, multi-storey frames would consist of individual programs (or groups of sub-programs if this arrangement was found to be more convenient), each concerned with a specific area of the problem. These areas are:

1. Slab analysis and design;
2. Frame analysis;
3. Beam design;
4. Column design;
5. A criterion for design acceptance which may not have a direct structural significance.

In addition to individual program or sub-program groups, if within one or more areas the designer is to be offered a choice of *method* then the number of available programs in the library will increase accordingly. Where a choice of method exists, and an option at one stage is compatible with more than one of the options at a later stage, then the freedom to pursue one route in preference to another is simplified if the input and output of information is, as far as possible, standardized.

Whilst the overall design operation is decision design based and the designer should be free to dictate the order in which the various stages are executed and even the degree of accuracy to which the results are produced, individual programs may owe more to automatic design principles than to those of decision design. This is a consideration for which there can be no rigid rules. The programmer's main concern should be to produce a flexible design tool and to achieve this he may have to allow the user's interests to take precedence over his own programming preferences.

5.3.2 Sequence of Decision Design Operations

The flow diagram shown in Fig. 5.3 outlines a suggested decision design procedure. In this diagram the 'balloons' represent stages in the design process rather than individual blocks of program.

Stage (A) is a one-off operation which lies outwith the frame design loop. It is convenient to analyse and proportion the slabs using a single program based upon automatic design principles. If it is intended to explore a variety of structural arrangements then the initial output from the slab design program should be restricted to a list of overall slab depths; it can be arranged for more detailed information to be output on rerunning the program after the final structural form has been decided.

The frame design operation is defined by a loop $B \rightarrow C \rightarrow D$. This is generated by the designer as he makes decisions and calls the analysis and member

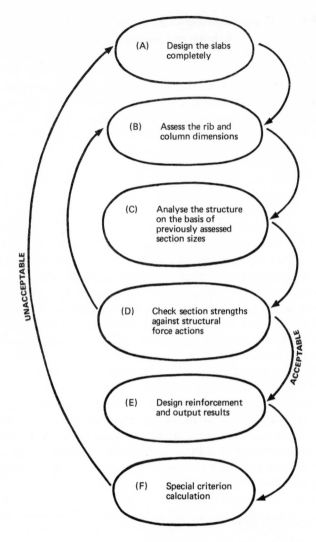

Figure 5.3 Flow diagram showing the basic areas in a
decision design library for the design of reinforced concrete
frames

design programs in the required order. To a large extent the time taken to complete one iteration will depend upon the facility with which the designer is able to make reasonable engineering decisions. The rib and column dimensions are assessed at Stage (B). These assessments may either be based upon an informed guess or the use of acceptable empirical methods. Since one purpose of computer aided design is to relieve the designer of tedious arithmetic, in the latter case a simple program based upon unsophisticated assessments would assist him in arriving at reasonable initial section sizes. For beams such a program could be based upon considerations of span/depth ratio together with an acceptable level of shear stress; columns might be assessed on the basis of their maximum allowable slenderness ratio and the approximate maximum axial loads they are to carry. In the interests of effectively reducing the number of design iterations it could even be profitable to develop a simple automatic design based program of the calibre of RCF12 (see Section 5.2) to provide an even finer initial estimate.

The assumed section dimensions together with details of the structural geometry and loading would comprise the data for the first and subsequent frame analyses at Stage (C). If, as is the case with this type of design operation, the analysis program is to be rerun a number of times it is expedient to build data which is common to all analyses into the program and to read in each time only that part of the information which represents the modified sections.

At the conclusion to an interim analysis the output of information should be restricted to what is immediately relevant—for example, the force actions and displacements that are necessary for a check to be made upon key members chosen by the designer as being typical of groups of members. Since it is probable that all the spans in a continuous beam would have the same rib dimensions it is only necessary to know the maximum value of each force action type (shear force, support and span bending moment) occurring within the whole length of the beam in order to assess the suitability of its overall dimensions. Similar restrictions apply to the output of maximum force action combinations occurring within groups of columns at one level.

The suitability of the assumed sections may now be checked at Stage (D). Again it is advisable to restrict the output of design information at an interim stage. Working with the overall concrete sizes and the maximum force actions as data, all that is necessary of an interim design check is an indication of reinforcement percentages required in representative members. This information is sufficient to allow a designer to assess their suitability. If at this stage one or more of the assumed sections prove to be unacceptable then the $B \to C \to D$ loop may be recycled on the basis of updated information. If section modifications are only marginal then it may not be considered necessary to reactivate Stage (C). In this matter the designer will exercise his own judgement. The iteration procedure would be continued until the designer was satisfied that he had an acceptable solution.

Unless datafile facilities are available it will be necessary to carry out a final analysis of the structure once all the key members had been approved in order to obtain a complete record of the design forces and displacements. The beam and column design programs are called for the last time at Stage (E). Their output should now be switched to give formal reinforcement arrangements. The amount of information given by a design program is obviously governed by the facilities which have been built into it. These can range from simple reinforcement arrangements at critical sections to complete designs which include detail drawings and bar bending schedules. But since such massive amounts of information are expensive to produce in terms of both time and resources its output would naturally be suppressed until the final form of the design had been decided upon.

Ideally the design process is one which leads to a 'best' solution in terms of cost, economical use of materials or some other criterion. A suitable group of programs could be developed for such assessments to be made at Stage (F). An optimum solution would result from varying the design parameters and generating the loop $A \rightarrow B \rightarrow C \rightarrow D \rightarrow E \rightarrow F$ as often as is necessary to achieve the desired result.

5.4 The 'All-embracing' Program—An Alternative Decision Design Approach to Reinforced Concrete Frames

5.4.1 Program Specification—RCF25

The purpose in presenting this program is to demonstrate that the same flexibility of choice may be built into an 'all-embracing' decision design program as is available to the user of an equivalent design library.

The automatic design based Program RCF12 provided the framework from which RCF25 was created. Because of this there are extensive areas common to both programs—notably those concerned with the handling of information, the design of slabs and the frame analysis. With one exception the limitations and lack of sophistication of Program RCF12 are also to be found in RCF25. The exception, a direct consequence of the basically different design approaches offered by the two programs, is the way in which the beams and columns are designed. In the case of RCF12 the automatic design of members is possible only if the solution is anticipated at the outset by defining the relative dimensions of each section and the reinforcement percentages. Such constraints lead to a unique section for each combination of force actions. In contrast with this the decision design methods of Program RCF25 allow the designer as much freedom in section choice as he would have in a similar manual design context. The only constraints which are offered are those of acceptable practice, and to this end the criteria built into the program will automatically cause sections to be rejected whose calculated reinforcement lies outwith acceptable maximum or minimum percentages.

5.4.2 The RCF25 Flow Diagram

The flow diagram for this program is shown in Fig. 5.4. Since the basic design areas are similar to those already discussed in Sections 5.2 and 5.3 they will not be further elaborated upon here. It is of more interest to look at the order in which designs are executed and the consequences (in programming terms) of making one decision in preference to another.

Briefly, control is exercised by the use of integer variables U and X as triggers to initiate certain actions automatically, together with five 'gates' (labelled I to V in the flow diagram) which allow the designer to follow a variety of paths through the design procedure.

Two different forms of results output are available—*provisional results* and *full results*. Provisional results are wholly diagnostic. They comprise a list of those sections which were modified during the latest design attempt (and as will be seen, this does not necessarily mean *all* the representative sections) together with a simple diagnosis of their suitability. Full results on the other hand always comprise the whole list of representative sections together with their steel percentages if the sections are suitable according to the programmed criteria and diagnostic comments if they are not.

For a first design attempt the variables U and X are given programmed values of 0 and 1 respectively outwith the main design loops. The role of U is to select the correct section input headings according to whether it is a first or a subsequent design attempt. U takes the value of 1 after the initial section input and thereafter the section input requirements automatically change. Whenever X = 1 (sometimes programmed and at other times the result of a designer decision at Gate I) the design procedure is channelled through all of the major design areas—analysis, determination of design force actions and calculation of reinforcement. At the designer's discretion a Gate I choice of X = 0 (and this may only be made after the first design attempt) bypasses the analysis procedure and assesses modified sections on the basis of force actions resulting from a previous analysis. This is a useful time saving device if in the designer's view the latest round of modifications do not merit a new analysis. In such cases only the modified sections are redesigned and this situation is treated by the program as a tentative design for which only provisional results can be obtained. Thus provisional results are only concerned with newly designed sections which may or may not have been based upon a current analysis. Full results accurately summarize the state of a design and may therefore only follow in the wake of a complete design procedure.

A first design attempt (U = 0, X = 1) bypasses Gate I and follows the whole design procedure. Because the results of this are the product of a fresh analysis the designer is given the choice at Gate II between electing for provisional or full results. If full results are chosen then the program run may either be terminated at Gate III or continued with the input of modified section information. In the latter case U now becomes equal to 1 and the computer therefore 'knows' that henceforth it will be dealing with modifications to the original design. If provisional results were chosen at Gate II then Gate IV offers the designer a choice between following them with a full set of results or of modifying sections; in either case U is set equal to 1. Because U is now permanently equal to 1 a halt will always be made at Gate I. This offers the designer a choice between designing the modified sections on the basis of force actions retrieved from the previous analysis (X = 0) or of following through the whole design procedure (X = 1). In the latter case an interim design will follow along the lines of a first attempt. But if X is equal to 0 then only Provisional results are possible and these are produced automatically, followed by a halt at Gate V.

178

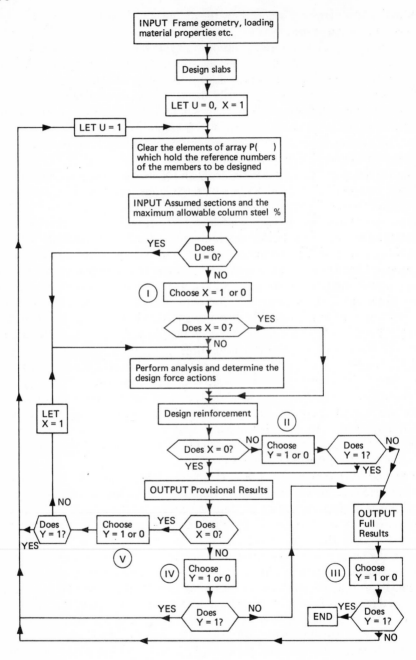

Figure 5.4 Flow diagram for program RCF25 (the decision design of reinforced concrete frames)

The choice now facing the designer is either to accept the provisional results tentatively (in which case X takes a programmed value of 1, the current form of the structure is analysed and all the sections are assessed on the basis of this analysis), or alternatively he may elect to modify sections and return to a Gate I decision.

Whilst manipulating the program the designer need not consciously think of 'gates' and the further consequences of the decisions he takes. In the natural course of a design calculation the program will halt and present him with an either/or situation. Depending upon his response the design sequence compatible with that decision follows automatically.

5.4.3 Description of Program RCF25

A list of the variables and arrays used in this program is given in Section 5.2.3. The following RCF25 program areas are of identical form to those in RCF12, but because the two programs are of different length and construction the statement numbers do not match. They have already been discussed in Section 5.2 and are listed below without further comment.

Statements 100–410:   Input frame geometry, loading, etc.
Statements 420–560:   Slab design.
Statements 570–650:   Compute minimum allowable section dimensions.
Statements 1320–1450: Compute beam and column second moments of area.
Statements 1460–1480: Compute beam design loads.
Statements 1490–1560: Compute fixed end moments.
Statements 1570–1720: Compute member stiffnesses.
Statements 1730–1910: Moment distribution procedure.
Statements 1950–2000: Compute beam shears.
Statements 2220–2280: Determine the maximum moment and corresponding axial load in external columns.

The program is listed below and should be read in conjunction with the flow diagram which is shown in Fig. 5.4.

```
100 PRINT "NUMBER OF BAYS AND STOREYS=";
110 INPUT B,S
120 PRINT "NOTE - ALL FRAME BEAM SPANS MUST BE EQUAL"
130 PRINT "SPAN (M)=";
140 INPUT L
150 PRINT "STOREY HEIGHTS (M)"
160 FOR I=1 TO S
170 IF I>1 THEN 200
180 PRINT "  TOP STOREY HEIGHT=";
190 GOTO 210
200 PRINT "HEIGHT OF STOREY"S-I+1"=";
210 INPUT S(I)
220 NEXT I
230 PRINT "DISTANCE BETWEEN FRAMES (M)=";
240 INPUT F1
250 PRINT "MINIMUM SLAB DEPTH (MM)=";
260 INPUT S1
```

```
270 PRINT "CHARACTERISTIC IMPOSED LOADS (KN/M+2)"
280 PRINT "AT ROOF AND FLOOR LEVELS"
290 FOR I=1 TO S
300 IF I>1 THEN 330
310 PRINT "          ROOF LOAD=";
320 GOTO 340
330 PRINT "LOAD AT FLOOR"S-I+1"=";
340 INPUT W(I)
350 NEXT I
360 PRINT "MAXIMUM COLUMN STEEL PERCENTAGE=";
370 INPUT R3
380 PRINT "FCU AND FY=";
390 INPUT F2,F3
400 PRINT "NUMBER OF INDIVIDUAL JOINT BALANCES/ITERATION=";
410 INPUT G1
420 FOR I=1 TO S
430 IF F1*1000/26+30>S1 THEN 460
440 T(I)=S1
450 GOTO 470
460 T(I)=F1*1000/26+30
470 W1=T(I)*24/1000
480 V(I)=1.4*W1+1.6*W(I)
490 M1=V(I)*F1+2/9
500 M2=0.15*F2*(T(I)-30)+2/1000
510 IF M2>=M1 THEN 540
520 T(I)=T(I)+10
530 GOTO 470
540 IF T(I)/10=INT(T(I)/10) THEN 560
550 T(I)=INT(T(I)/10)*10+10
560 NEXT I
570 PRINT
580 PRINT
590 J1=INT((L*1000/26-T(I))/20)*20+80
600 J2=INT(S(S)*1000/(60*20))*20+20
610 PRINT "NOTE - MINIMUM ALLOWABLE RIB DEPTH="J1"MM"
620 PRINT
630 PRINT "        MINIMUM ALLOWABLE COLUMN SECTION DIMENSION="J2"MM"
640 PRINT
650 PRINT
660 U=0
670 X=1
680 FOR I=1 TO S
690 FOR J=B+2 TO B+4
700 P(I,J)=0
710 NEXT J
720 NEXT I
730 IF U=0 THEN 790
740 PRINT "HOW MANY BEAMS ARE TO BE MODIFIED";
750 INPUT X6
760 PRINT
770 IF X6=0 THEN 920
780 GOTO 830
790 PRINT "INPUT INITIAL BEAM DIMENSIONS"
800  PRINT "----- ----------------------"
810 PRINT
820 X6=S
830 PRINT "SPECIFY FLOOR LEVEL,RIB DEPTH (MM) AND RIB WIDTH (MM)"
840 FOR I=1 TO X6
850 INPUT X9,X8,X7
860 D(S-X9+1)=X8+T(S-X9+1)
870 B(S-X9+1)=X7
880 P(S-X9+1,B+2)=S-X9+1
890 G(S-X9+1)=X8
900 NEXT I
910 PRINT
920 IF U=0 THEN 980
```

```
930 PRINT "HOW MANY EXTERNAL COLUMNS ARE TO BE MODIFIED";
940 INPUT X6
950 PRINT
960 IF X6=0 THEN 1100
970 GOTO 1020
980 PRINT "INPUT INITIAL EXTERNAL COLUMN DIMENSIONS"
990 PRINT "-------------------- -----------------"
1000 PRINT
1010 X6=S
1020 PRINT "SPECIFY STOREY COLUMN DEPTH (MM) AND COLUMN WIDTH (MM)"
1030 FOR I=1 TO X6
1040 INPUT X9,X8,X7
1050 C(S-X9+1,1)=X8
1060 A(S-X9+1,1)=X7
1070 P(S-X9+1,B+3)=S-X9+1
1080 NEXT I
1090 PRINT
1100 IF B=1 THEN 1270
1110 IF U=0 THEN 1160
1120 PRINT "HOW MANY INTERNAL COLUMNS ARE TO BE MODIFIED";
1130 INPUT X6
1140 IF X6=0 THEN 1280
1150 GOTO 1200
1160 PRINT "INPUT INITIAL INTERNAL COLUMN DIMENSIONS"
1170 PRINT "----- ------- -------------------------"
1180 PRINT
1190 X6=S
1200 PRINT "SPECIFY STOREY COLUMN DEPTH (MM) AND COLUMN WIDTH (MM)"
1210 FOR I=1 TO X6
1220 INPUT X9,X8,X7
1230 C(S-X9+1,2)=X8
1240 A(S-X9+1,2)=X7
1250 P(S-X9+1,B+4)=S-X9+1
1260 NEXT I
1270 PRINT
1280 IF U=0 THEN 1320
1290 PRINT "IF FRAME TO BE REANALYSED TYPE 1 ELSE 0";
1300 INPUT X
1310 IF X=0 THEN 2290
1320 FOR I=1 TO S
1330 E(I)=140*L+B(I)
1340 IF E(I)<1000*F1 THEN 1360
1350 E(I)=1000*F1
1360 X1=((E(I)-B(I))*T(I)+2/2+B(I)*D(I)+2/2)
1361 X1=X1/((E(I)-B(I))*T(I)+B(I)*D(I))
1370 I(I)=(E(I)-B(I))*T(I)+3/12+B(I)*D(I)+3/12
1380 I(I)=I(I)+(E(I)-B(I))*T(I)*(X1-T(I)/2)+2
1390 I(I)=I(I)+B(I)*D(I)*(X1-D(I)/2)+2
1400 NEXT I
1410 FOR I=1 TO S
1420 FOR R=1 TO 2
1430 J(I,R)=A(I,R)*C(I,R)+3/12
1810 R(I,J)=R(I,J)-M3*N(I,J)/S2
1820 IF I=1 THEN 1840
1830 P(I-1,J)=P(I-1,J)-M3*K(I,J)/(2*S2)
1840 O(I+1,J)=O(I+1,J)-M3*L(I,J)/(2*S2)
1850 IF J=B+1 THEN 1880
1860 R(I,J+1)=R(I,J+1)-M3*M(I,J)/(2*S2)
1870 IF J=1 THEN 1890
1880 Q(I,J-1)=Q(I,J-1)-M3*N(I,J)/(2*S2)
1890 NEXT J
1900 NEXT I
1910 NEXT K
1920 FOR I=1 TO S
1930 F(I)=ABS(R(I,2))
1940 NEXT I
```

```
1950 FOR I=1 TO S
1960 FOR J=1 TO 2
1970 X(I,J)=-(Q(I,J)+R(I,J+1))/L+Y(I)*L/2
1980 Z(I,J+1)=(Q(I,J)+R(I,J+1))/L+Y(I)*L/2
1990 NEXT J
2000 NEXT I
2010 FOR I=1 TO S
2020 FOR J=1 TO 2
2030 IF J=1 THEN 2060
2040 R=2
2050 GOTO 2070
2060 R=1
2070 IF I>1 THEN 2100
2080 H(1,J)=X(1,J)+Z(1,J)+1.4*S(1)*A(1,R)*C(1,R)*24/1000000
2090 IF I=1 THEN 2110
2100 H(I,J)=H(I-1,J)+X(I,J)+Z(I,J)+1.4*S(I)*A(I,R)*C(I,R)*24/1000000
2110 NEXT J
2120 NEXT I
2130 IF B=1 THEN 2220
2140 FOR I=1 TO S
2150 IF ABS(P(I,2))<ABS(O(I+1,2)) THEN 2190
2160 U(I,4)=ABS(P(I,2))
2170 GOTO 2200
2190 U(I,4)=ABS(O(I+1,2))
2200 U(I,3)=H(I,2)
2210 NEXT I
2220 FOR I=1 TO S
2230 IF ABS(P(I,1))<ABS(O(I+1,1)) THEN 2260
2240 U(I,6)=ABS(P(I,1))
2250 GOTO 2270
2260 U(I,6)=ABS(O(I+1,1))
2270 U(I,5)=H(I,1)
2280 NEXT I
2290 FOR I=1 TO S
2300 IF X=1 THEN 2320
2310 IF P(I,B+2)=0 THEN 2430
2320 M5=0.15*F2*B(I)*0.81*D(I)+2/1000000
2330 IF M5<F(I) THEN 2390
2340 W=100*F2/(1.1*F3)
2350 Y=F(I)*F2*10+10/(0.957*F3+2*B(I)*0.81*D(I)+2)
2360 Q(I,B+2)=(W-SQR(W+2-4*Y))/2
2370 Q(I,B+3)=0
2380 GOTO 2430
2390 Z=(F(I)-M5)*10+6/(0.72*F3*0.8*D(I))
2400 Q(I,B+3)=Z*100/(B(I)*0.9*D(I))
2410 Z1=(0.2*F2*B(I)*0.9*D(I)+0.72*F3*Z)/(0.87*F3)
2420 Q(I,B+2)=Z1*100/(B(I)*0.9*D(I))
2430 NEXT I
2440 FOR R=1 TO 2
2450 IF R=2 THEN 2490
2460 P=6
2470 Q=5
2480 GOTO 2510
2490 P=4
2500 Q=3
2510 FOR I=1 TO S
2520 R1=R3
2530 IF X=1 THEN 2550
2540 IF P(I,B+2+R)=0 THEN 2870
2550 N3=U(I,Q)*1000/(F2*A(I,R)*C(I,R))
2560 N1=0.4+0.72*R1*F3/(100*F2)
2570 IF N1>N3 THEN 2630
2580 IF R1<R3 THEN 2610
2590 P(I,B+R+4)=1000
2600 GOTO 2870
2610 P(I,B+R+4)=R1+0.25
```

```
2620 GOTO 2870
2630 N2=0.18-0.00075*R1*F3/F2
2640 M8=0.063+0.00318*R1*F3/F2
2650 M9=0.003555*R1*F3/F2
2660 M7=U(I,Q)*1000*(0.75*S(I)*1000)+2/(1750*C(I,R))
2670 M7=M7*(1-0.0035*0.75*S(I)*1000/C(I,R))
2680 M7=(M7+U(I,P)*1000000)/(F2*A(I,R)*C(I,R)+2)
2690 IF N3>N2 THEN 2770
2700 A5=(M8-M9)*N3/N2+M9
2710 IF A5>M7 THEN 2840
2720 IF R1<R3 THEN 2750
2730 P(I,B+R+4)=1000
2740 GOTO 2870
2750 P(I,B+R+4)=R1+0.25
2760 GOTO 2870
2770 A5=(N1-N3)*M8/(N1-N2)
2780 IF A5>M7 THEN 2840
2790 IF R1<R3 THEN 2820
2800 P(I,B+R+4)=1000
2810 GOTO 2870
2820 P(I,B+R+4)=R1+0.25
2830 GOTO 2870
2840 R1=R1-0.25
2850 IF R1>1.0 THEN 2550
2860 P(I,B+R+4)=1.0
2870 NEXT I
2880 IF B=1 THEN 2900
2890 NEXT R
2900 IF X=0 THEN 2970
2910 PRINT
2920 PRINT
2930 PRINT "FOR PROVISIONAL RESULTS OR FULL RESULTS TYPE 1 OR 0";
2940 INPUT Y
2950 PRINT
2960 IF Y=0 THEN 3810
2970 PRINT
2980 PRINT "****************************************************"
2990 PRINT "* **********************************************  *"
3000 PRINT "* *                                            * *"
3010 PRINT "* *          PROVISIONAL DESIGN RESULTS        * *"
3020 PRINT "***          ---------- ------ -------         ***"
3030 FOR I=1 TO S
3040 IF P(I,B+2)=0 THEN 3180
3050 IF Q(I,B+2)>4.0 THEN 3110
3060 IF Q(I,B+3)>Q(I,B+2) THEN 3110
3070 IF Q(I,B+2)<=0.25 THEN 3150
3080 PRINT
3090 PRINT"BEAM"S-I+1"***CURRENT SECTION DIMENSIONS ACCEPTABLE***"
3100 GOTO 3180
3110 PRINT
3120 PRINT "BEAM"S-I+1"INCREASE SECTION DIMENSIONS CURRENTLY"
3130 PRINT "ESTIMATED AT D="G(I)"MM   B="B(I)"MM"
3140 GOTO 3180
3150 PRINT
3160 PRINT "BEAM"S-I+1"DECREASE SECTION DIMENSIONS CURRENTLY"
3170 PRINT "ESTIMATED AT D="G(I)"MM   B="B(I)"MM"
3180 NEXT I
3190 PRINT
3200 PRINT
3210 FOR R=1 TO 2
3220 IF R=2 THEN 3310
3230 D=0
3240 FOR I=1 TO S
3250 D=D+P(I,B+3)
3260 NEXT I
3270 IF D=0 THEN 3540
```

```
3280 PRINT "EXTERNAL COLUMNS"
3290 PRINT "-------- ------"
3300 GOTO 3390
3310 PRINT
3320 D=0
3330 FOR I=1 TO S
3340 D=D+P(I,B+4)
3350 NEXT I
3360 IF D=0 THEN 3540
3370 PRINT "INTERNAL COLUMNS"
3380 PRINT "-------- ------"
3390 FOR I=1 TO S
3400 IF P(I,B+R+4)=1.0 THEN 3500
3410 IF P(I,B+R+4)=1000 THEN 3460
3420 IF P(I,B+R+2)=0 THEN 3530
3430 PRINT
3440 PRINT"STOREY"S-I+1"COLUMN***CURRENT DIMENSIONS ACCEPTABLE***"
3450 GOTO 3530
3460 PRINT
3470 PRINT "STOREY"S-I+1"COLUMN - INCREASE SECTION DIMENSICNS"
3480 PRINT "CURRENTLY ESTIMATED AT D="C(I,R)"MM  B="A(I,R)"MM"
3490 GOTO 3530
3500 PRINT
3510 PRINT "STOREY"S-I+1"COLUMN - DECREASE SECTION DIMENSICNS"
3520 PRINT "CURRENTLY ESTIMATED AT D="C(I,R)"MM  B="A(I,R)"MM"
3530 NEXT I
3540 NEXT R
3550 PRINT "***                                            ***"
3560 PRINT "* *                                            * *"
3570 PRINT "* *                                            * *"
3580 PRINT "* ***************************************** *"
3590 PRINT "*******************************************"
3600 PRINT
3610 IF X=1 THEN 3740
3620 PRINT
3630 PRINT
3640 PRINT "IF SOME SECTIONS ARE STILL TO BE MODIFIED TYPE 1 ELSE 0";
3650 INPUT Y
3660 PRINT
3670 IF Y=0 THEN 3700
3680 U=1
3690 GOTO 680
3700 X=1
3710 GOTO 1320
3720 PRINT
3730 PRINT
3740 PRINT "EITHER - TYPE 1 - AND MODIFY SECTIONS"
3750 PRINT "    OR - TYPE 0 - TO OBTAIN FULL PRINTOUT OF"
3751 PRINT "                  CURRENT RESULTS."
3760 INPUT Y
3770 PRINT
3780 IF Y=0 THEN 3810
3790 U=1
3800 GOTO 680
3810 PRINT
3820 PRINT
3830 PRINT "*******************************************"
3840 PRINT "* ***************************************** *"
3850 PRINT "* *                                            * *"
3860 PRINT "* *          CURRENT SECTION INFORMATION       * *"
3870 PRINT "* *          ------- ------- -----------       * *"
3880 PRINT "***                                            ***"
3890 FOR I=1 TO S
3900 IF I>1 THEN 3930
3910 PRINT "        ROOF SLAB DEPTH="T(I)"MM"
```

```
3920 GOTO 3940
3930 PRINT "SLAB DEPTH AT FLOOR"S-I+1"="T(I)"MM"
3940 NEXT I
3950 PRINT
3960 PRINT "RIB DIMENSIONS"
3970 PRINT "--- ----------"
3980 PRINT
3990 FOR I=1 TO S
4000 IF I>1 THEN 4070
4010 PRINT "   ROOF LEVEL RIB : D="G(I)"MM    B="B(I)"MM"
4020 IF Q(I,B+2)>4.0 THEN 4050
4030 PRINT "                  AST="Q(I,B+2)"%    ASC="Q(I,B+3)"%"
4040 GOTO 4120
4050 PRINT "                  AST>MAXIMUM ALLOWABLE PERCENTAGE"
4060 GOTO 4120
4070 PRINT "FLOOR"S-I+1"LEVEL RIB : D="G(I)"MM    B="B(I)"MM"
4080 IF Q(I,B+2)>4.0 THEN 4110
4090 PRINT "                  AST="Q(I,B+2)"%    ASC="Q(I,B+3)"%"
4100 GOTO 4120
4110 PRINT "                  AST>MAXIMUM ALLOWABLE PERCENTAGE"
4120 NEXT I
4130 PRINT
4140 FOR R=1 TO 2
4150 IF R=1 THEN 4210
4160 PRINT
4170 PRINT "INTERNAL COLUMN DIMENSIONS"
4180 PRINT "-------- ------ ----------"
4190 PRINT
4200 GOTO 4240
4210 PRINT "EXTERNAL COLUMN DIMENSIONS"
4220 PRINT "-------- ------ ----------"
4230 PRINT
4240 FOR I=1 TO S
4250 IF I>1 THEN 4330
4260 PRINT "TOP STOREY COLUMNS : D="C(I,R)"MM    B="A(I,R)"MM"
4270 IF P(I,B+R+4)>R3 THEN 4300
4280 PRINT "                  ASC="P(I,B+R+4)"%"
4290 GOTO 4370
4300 PRINT "                  ASC>MAXIMUM ALLOWABLE PERCENTAGE"
4310 GOTO 4370
4320 IF P(I,B+R+4)>R3 THEN 4350
4330 PRINT " STOREY"S-I+1"COLUMNS : D="C(I,R)"MM    B="A(I,R)"MM"
4335 IF P(I,B+R+4)>R3 THEN 4350
4340 GOTO 4360
4350 PRINT "                  ASC>MAXIMUM ALLOWABLE PERCENTAGE"
4355 GOTO 4370
4360 PRINT "                  ASC="P(I,B+R+4)"%"
4370 NEXT I
4380 IF B=1 THEN 4400
4390 NEXT R
4400 PRINT "***                                      ***"
4410 PRINT "* *                                      * *"
4420 PRINT "* ***********************************************  **"
4430 PRINT "***********************************************  **"
4440 PRINT
4450 PRINT
4460 PRINT "IF RESULTS ARE ACCEPTABLE TYPE 1 ELSE 0";
4470 INPUT Y
4480 PRINT
4490 PRINT
4500 IF Y=1 THEN 4530
4510 U=1
4520 GOTO 680
4530 END
```

5.4.3.1 Statements 660 and 670

The integer variables U and X are set to the values of 0 and 1 respectively. The fact that $U = 0$ informs the computer that this is the first of a series of trial designs. When $X = 1$ it ensures that the proposed structure which represents this attempt will be analysed before the sections are proportioned.

5.4.3.2 Statements 680–720

Prior to an input of section dimensions the contents of columns $(B + 2)$, $(B + 3)$ and $(B + 4)$ in array P( ) are set to zero. Subsequently the contents of some or all of these elements become non-zero. The role of these elements is to act as triggers which control the number of sections to be designed if at an interim design stage it is decided to design sections on the basis of an out-dated analysis.

5.4.3.3 Statements 730–1270

This block of statements is concerned with the input of the proposed rib and column section dimensions. The discussion will be confined to rib dimension input since this procedure is typical of the columns as well.

The procedure is arranged to meet the different input requirements of both the initial and the interim design attempts. For an initial design attempt $U = 0$. This occasions a jump from line 730 to line 790 where the appropriate command INPUT INITIAL BEAM DIMENSIONS is printed out. The content of X6 represents the number of beams to be designed and since the first attempt requires that all the beams will be designed then this variable is automatically set equal to S (the total number of beams) at line 820.

Prior to subsequent design attempts U is set equal to 1 elsewhere in the program and when line 730 is again encountered the jump to line 790 is ignored. Instead, the question HOW MANY BEAMS ARE TO BE MODIFIED? is printed out at line 740. The designer is then able to assign a value between 0 and S to X6 at line 750. If $X6 = 0$ then the program jumps to line 920 where a request for external column design information is made. Otherwise a non-zero content in X6 restricts the amount of section information which needs to be input at an interim design stage (see the limit to the value of I at line 840).

The procedure for storing section dimensions is identical whatever the design stage. The content of X6 governs the number of times (see line 840) that information is requested. Each input consists of three pieces of data (the floor level, the rib depth and the rib width) and these are read into the variables X9, X8 and X7 (in that order) at line 850. By defining a floor level in the usual way (i.e. 1st, 2nd, 3rd, etc. above Ground Floor Level) the relevant row for that section information in the arrays D( ), B( ) and G( ) is given by $(S - X9 + 1)$—see lines 860 to 890. The fact that $P(S - X9 + 1, B + 2)$ takes a non-zero value at line 880 is significant at a later stage in the program when the contents of the elements in column $(B + 2)$ of array $P(\ )$ determine which beam sections will be designed.

5.4.3.4 Statements 1280–1310

These statements comprise Gate I. Here the designer is presented with a choice between two courses of action. It might be profitable to amplify a PRINT statement of the type given at line 1290 with further statements describing the consequences of each action. However, in the interests of rapid decision making it should be possible to suppress this extra printout once the program user is familiar with its message.

The simplest form of gate is that used in Program RCF25. In general any number of options may be handled by identifying a specific course of action with a unique reference number.

Other gates in this program occur at the following groups of statements:

Gate II–Statements 2930–2960
Gate III–Statements 4460–4500
Gate IV–Statements 3740–3780
Gate V–Statements 3640–3670

5.4.3.5 Statements 1920–1940

In contrast with RCF12 no search is made in this program to find the maximum beam support moment at each floor level. It is correctly assumed to occur at the right hand end of the first span in each continuous beam.

5.4.3.6 Statements 2010–2210

This block of statements is concerned with determining column design force action combinations. Because of the restrictions placed upon the type of structure and its loading the second (and penultimate) lines of columns are the most heavily loaded, and are also subjected to the greatest bending moments, of all the internal columns. By accepting this fact there is no need to program a search for maximum force action combinations.

The axial loads which occur in the first two lines of columns are calculated at lines 2010 to 2120 and stored for reference in array $H(\ )$. For each lift of the first line of internal columns the maximum moment in that column is paired with the appropriate axial load (see lines 2130 to 2210) and stored in the 4th and 3rd columns of array $U(\ )$ respectively.

5.4.3.7 Statements 2290–2430

The purpose of these statements is to design the beams to meet the requirements of bending at the first internal support. In this program the T-beam span design has been omitted since it is unlikely that within the present constraints of structural type and loading the span moment would influence beam size.

If X is equal to 1 then according to the statement at line 2300 all of the beams will be designed. Otherwise, for a beam to be considered, $P(I, B + 2)$ must have a non-zero value (see Sections 5.4.3.2 and 5.4.3.3).

The concrete moment of resistance for 'balanced' conditions and an effective depth equal to ($0.9 \times$ the overall depth), i.e. M5, is calculated at line 2320. This value is compared with the applied moment $F(I)$ at line 2330. If M5 is greater or equal to $F(I)$ then the steel calculation will be one for a section having single reinforcement. This takes place at lines 2340 to 2360. The beam tension steel percentage is stored in $Q(I, B + 2)$ and the percentage of compression steel (in this case zero) in $Q(I, B + 3)$.

If M5 is less than $F(I)$ then a jump to line 2390 recognizes the fact that the section should be doubly reinforced. The required percentages of compression and tension steel are recorded in $Q(I, B + 3)$ and $Q(I, B + 2)$ respectively (see lines 2400 and 2420).

At this stage the steel percentages are stored without comment. It is not until results are required that these values are tested against programmed criteria to determine whether the assumed beam sections are indeed acceptable.

5.4.3.8 Statements 2440–2890

The purpose of this block of statements is to design the columns. The value which is assigned to R at line 2440 determines whether the external or the internal columns are to be designed. As a result of this the values taken by P and Q then ensure that the relevant force action combinations are chosen (see Section 5.2.3.8).

In the column design procedure the contents of X and $P(I, B + 2 + R)$ (see lines 2530 and 2540) play the same role in determining whether a given column should be designed as did X and $P(I, B + 2)$ in the beam design situation.

In this program the columns are designed to fulfil the requirements of the same column design envelope (see Fig. 5.2) as were the columns in RCF12. However, the two design approaches differ in one essential respect; in RCF12 the steel percentage was specified, and following this a concrete section was found to meet this requirement—in contrast with the current problem where the concrete dimensions are known and the amount of steel must be calculated.

In order to achieve designs which give realistic amounts of steel an acceptable upper limit to the steel percentage (R3) is specified by the designer at the information input stage (see lines 360 and 370). A lower limit of 1% is specified within the program at lines 2850 and 2860.

The flow diagram which is shown in Fig. 5.5 will assist in interpreting the column design procedure. Each item in the flow diagram is related to specific line numbers in the procedure.

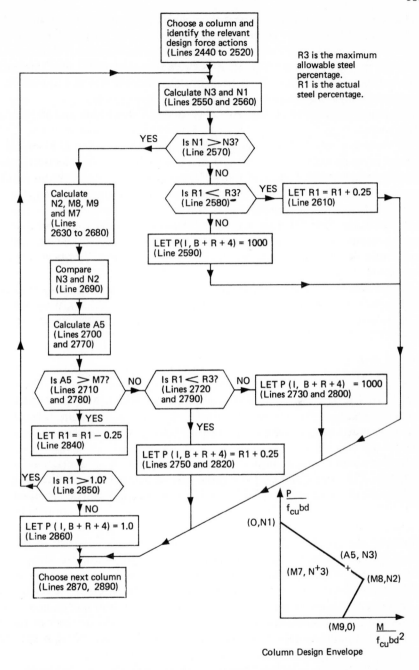

Figure 5.5 Column design flow diagram—lines 2440 to 2890 of program RCF 25

The design of a column section begins with the assumption that it is reinforced with its maximum percentage of steel (see line 2520). A check at line 2570 determines whether the column strength at zero moment (N1) is greater than the design requirement (N3). If N1 is less than N3 then the column is understrength and at line 2590 the array element $P(I, B + R + 4)$, which is assigned to store the required steel percentage, is arbitrarily set to a value of 1000. This will trigger off the correct diagnostic response at the results output stage.

If, according to this first check, the column section does have a reserve of strength over that which is required, then a jump to line 2630 allows the procedure to continue—one which at this stage is substantially the same as that in RCF12 (see Section 5.2.3.8). The coordinates of a point defined by ($M7 = M/f_{cu} bd^2$, $N3 = P/f_{cu} bd$) are checked in relation to the column design envelope. If in the first instance the point lies outside the envelope then the section is too small to develop the combined force actions. Therefore $P(I, B + R + 4)$ is set to 1000 at line 2800. If the point does lie within the envelope then R1 is successively reduced by decrements of 0.25% (see line 2840) until one of two possible situations arises. Either R1 is reduced to below 1%, in which case $P(I, B + R + 4)$ is set to the value of 1% at line 2860—or at some steel percentage greater than 1% it is found that M7 is greater than A5 (see lines 2710 or 2780). The current value of $R1 + 0.25\%$ is then accepted at line 2820.

5.4.3.9 Statements 2980–3590

It is particularly important in decision design programming to present the results in as digestible a form as possible. All numerical results should be amplified by including sufficient descriptive printout to make their meaning obvious to the program user. Provisional results may, as is the case with RCF25, dispense with numerical answers completely and present instead a verbal interpretation of them. The sheer volume of output from decision design based programs makes it necessary to arrange for each parcel of results to stand out conspicuously from the remainder. In RCF25 this is achieved by using a title heading together with identifiable patterns of asterisks at the top and bottom of each output.

Provisional results play an advisory role. They indicate that a section may or may not be used but give no hard information concerning percentages of reinforcement. The word 'acceptable', used to indicate section approval in diagnostic messages, is a relative term; the designer may still modify such sections to meet his own standards.

The provisional result for a beam is only available if the content of $P(I, B + 2)$ is greater than zero (see Sections 5.4.3.2 and 5.4.3.3). A jump from line 3040 to line 3180 omits this result if $P(I, B + 2)$ is equal to zero. The beam tension and compression steel percentages are held in columns $(B + 2)$ and $(B + 3)$ of array $Q(\ )$. The statements at lines 3040 to 3060 introduce the steel percentage limits which can lead to the automatic rejection of a section at lines 3120 or

3160. Those sections with reinforcement percentages within these limits are accepted at line 3090.

Both the external and the internal column groups share a common output procedure between lines 3390 and 3530. Each group, however, needs a distinctive title (see lines 3280 and 3370). But the relevant title is suppressed if no column has been redesigned during the current trial and there is therefore no provisional result for that group. For external columns the check to ascertain whether there is a provisional result to print out is made at lines 3230 to 3270. A variable called D is initially set to zero at line 3230. Following this D takes the value of the sum of the contents of the elements in column $(B + 3)$ of array P( ). Non-zero elements in this column indicate that sections have been redesigned during the latest trial. Thus if D is greater than zero then the title heading at line 3280 is printed out, and a jump to line 3390 is followed by an output of provisional results. Otherwise, if D is equal to zero, the external columns are ignored. The criterion for an output of provisional internal column results is treated in a similar way at line 3320 to 3360.

5.4.3.10 Statements 3830–4440

Full results give an accurate summary (within the constraints set by the program) of the state of all the key sections in the structure. The steel percentages which are quoted are always derived from a design which is based upon a current analysis of the structure. Except for cases where the calculated steel percentages lie outwith the allowable range (in which case a message informs the designer of this fact) the designer is at liberty to accept or reject sections in the light of his own experience.

5.4.4 Example of Program RCF25 Output

A typical solution gives by this program appears at the end of this section. Following the input of data which described a 3-bay, 4-storey frame, its loading, material properties, etc., the computer responded with a note on the minimum possible section dimensions for this particular case.

The input of a full set of guessed section dimensions completed the description of the structure and caused the computer to execute the whole design procedure, at the end of which the designer was offered a choice between two kinds of results output. In this instance he opted for a set of provisional results. These informed him that the sections suggested for the first storey external columns, and the first and third storey internal columns, were undersized. The fact that the third and fourth storey external column sections had proved to be too large was discounted since it was felt that their dimensions were already at a reasonably practical minimum value. The remaining column sections were either of acceptable size, or too large—advice which had to be ignored since it was intended to pair them off with others (which had already proved to be too small) in 'same-section' groups.

The designer therefore elected to modify the undersized column sections together with those of the columns with which they were paired. The order of change involved justified a reanalysis of the structure. Following this a further set of provisional results was requested. This time the scope of the provisional results was automatically restricted a survey of those groups of members in which changes had been made. (It should be remembered that the force actions in the group of modified members are not the only ones to be affected by a round of section modifications. If it is suspected that initially acceptable sections from other representative groups might be affected to such a degree as to require subsequent change, then provisional information concerning the status of these members may be derived by giving the machine the *same* section to work with—in the guise of a modification—that it was dealing with previously. In the present case, for example, had the computer been informed that one beam was to be modified, and the initial roof beam section had been reinput, then the provisional results would have included a diagnostic comment on *all* the beams that had now failed to meet the new design criteria.)

The provisional review of column sections now showed that (within the limitations set by the designer) they were responding satisfactorily; the designer therefore elected to receive a full output of current results. This review indicated that since the beam support sections required no compression steel they might reasonably be decreased in depth. A further round of member modifications, reanalysis and provisional results was therefore initiated. This culminated in a set of section sizes which the designer elected to accept.

```
RCF25

NUMBER OF BAYS AND STOREYS= ? 3,4
NOTE - ALL FRAME BEAM SPANS MUST BE EQUAL
SPAN (M)= ? 7.4
STOREY HEIGHTS (M)
   TOP STOREY HEIGHT= ? 4.8
HEIGHT OF STOREY 3 = ? 4.8
HEIGHT OF STOREY 2 = ? 4.8
HEIGHT OF STOREY 1 = ? 5.7
DISTANCE BETWEEN FRAMES (M)= ? 3.6
MINIMUM SLAB DEPTH (MM)= ? 100
CHARACTERISTIC IMPOSED LOADS (KN/M↑2)
AT ROOF AND FLOOR LEVELS
        ROOF LOAD= ? 1.4
LOAD AT FLOOR 3 = ? 5.5
LOAD AT FLOOR 2 = ? 5.5
LOAD AT FLOOR 1 = ? 5.5
MAXIMUM COLUMN STEEL PERCENTAGE= ? 4.5
FCU AND FY= ? 25,250
NUMBER OF INDIVIDUAL JOINT BALANCES/ITERATION= ? 10


NOTE - MINIMUM ALLOWABLE RIB DEPTH= 180 MM

        MINIMUM ALLOWABLE COLUMN SECTION DIMENSION= 100 MM


INPUT INITIAL BEAM DIMENSIONS
----- -------- ---- -----------
```

```
SPECIFY FLOOR LEVEL, RIB DEPTH (MM) AND RIB WIDTH (MM)
 ? 4,350,250
 ? 3,500,250
 ? 2,500,250
 ? 1,500,250

INPUT INITIAL EXTERNAL COLUMN DIMENSIONS
----- ------- -------- ------ ----------

SPECIFY STO REY, COLUMN DEPTH (MM) AND COLUMN WIDTH (MM)
 ? 4,200,300
 ? 3,200,300
 ? 2,200,300
 ? 1,200,300

INPUT INITIAL INTERNAL COLUMN DIMENSIONS
----- ------- -------- ------ ----------

SPECIFY STO REY, COLUMN DEPTH (MM) AND COLUMN WIDTH (MM)
 ? 4,200,300
 ? 3,200,300
 ? 2,300,300
 ? 1,300,300


FOR PROVISIONAL RESULTS OR FULL RESULTS TYPE 1 OR 0 ? 1


****************************************************
* ************************************************* *
* *                                              * *
* *           PROVISIONAL DESIGN RESULTS         * *
***            ----------- ------ -------         ***

BEAM 4 ***CURRENT SECTION DIMENSIONS ACCEPTABLE***

BEAM 3 ***CURRENT SECTION DIMENSIONS ACCEPTABLE***

BEAM 2 ***CURRENT SECTION DIMENSIONS ACCEPTABLE***

BEAM 1 ***CURRENT SECTION DIMENSIONS ACCEPTABLE***


EXTERNAL COLUMNS
-------- -------

STOREY 4 COLUMN - DECREASE SECTION DIMENSIONS
CURRENTLY ESTIMATED AT D= 200 MM  B= 300 MM

STOREY 3 COLUMN - DECREASE SECTION DIMENSIONS
CURRENTLY ESTIMATED AT D= 200 MM  B= 300 MM

STOREY 2 COLUMN***CURRENT DIMENSIONS ACCEPTABLE***

STOREY 1 COLUMN - INCREASE SECTION DIMENSIONS
CURRENTLY ESTIMATED AT D= 200 MM  B= 300 MM

INTERNAL COLUMNS
-------- -------

STOREY 4 COLUMN - DECREASE SECTION DIMENSIONS
CURRENTLY ESTIMATED AT D= 200 MM  B= 300 MM

STOREY 3 COLUMN - INCREASE SECTION DIMENSIONS
CURRENTLY ESTIMATED AT D= 200 MM  B= 300 MM

STOREY 2 COLU MN***CURRENT DIMENSIONS ACCEPTABLE***
```

```
STOREY 1 COLUMN - INCREASE SECTION DIMENSIONS
CURRENTLY ESTIMATED AT D= 300 MM  B= 300 MM
***                                           ***
* *                                           * *
* *                                           * *
* ********************************************* *
*************************************************

EITHER - TYPE 1 - AND MODIFY SECTIONS
    OR - TYPE 0 - TO OBTAIN FULL PRINTOUT OF
                  CURRENT RESULTS.
 ? 1


HOW MANY BEAMS ARE TO BE MODIFIED ? 0

HOW MANY EXTERNAL COLUMNS ARE TO BE MODIFIED ? 2

SPECIFY STOREY, COLUMN DEPTH (MM) AND COLUMN WIDTH (MM)
 ? 2,250,300
 ? 1,250,300

HOW MANY INTERNAL COLUMNS ARE TO BE MODIFIED ? 4
SPECIFY STOREY, COLUMN DEPTH (MM) AND COLUMN WIDTH (MM)
 ? 4,250,300
 ? 3,250,300
 ? 2,450,300
 ? 1,450,300

IF FRAME TO BE REANALYSED TYPE 1 ELSE 0 ? 1


FOR PROVISIONAL RESULTS OR FULL RESULTS TYPE 1 OR 0 ? 1


*************************************************
* ********************************************* *
* *                                           * *
* *          PROVISIONAL DESIGN RESULTS        * *
***          ----------- ------ -------        ***


EXTERNAL COLUMNS
-------- -------

STOREY 4 COLUMN - DECREASE SECTION DIMENSIONS
CURRENTLY ESTIMATED AT D= 200 MM  B= 300 MM

STOREY 3 COLUMN - DECREASE SECTION DIMENSIONS
CURRENTLY ESTIMATED AT D= 200 MM  B= 300 MM

STOREY 2 COLUMN - DECREASE SECTION DIMENSIONS
CURRENTLY ESTIMATED AT D= 250 MM  B= 300 MM

STOREY 1 COLUMN***CURRENT DIMENSIONS ACCEPTABLE***

INTERNAL COLUMNS
-------- -------

STOREY 4 COLUMN - DECREASE SECTION DIMENSIONS
CURRENTLY ESTIMATED AT D= 250 MM  B= 300 MM

STOREY 3 COLUMN***CURRENT DIMENSIONS ACCEPTABLE***

STOREY 2 COLUMN - DECREASE SECTION DIMENSIONS
CURRENTLY ESTIMATED AT D= 450 MM  B= 300 MM
```

```
STOREY 1 COLUMN***CURRENT DIMENSIONS ACCEPTABLE***
***                                              ***
* *                                              * *
* *                                              * *
* ****************************************************** *
******************************************************

EITHER - TYPE 1 - AND MODIFY SECTIONS
   OR - TYPE 0 - TO OBTAIN FULL PRINTOUT OF
                 CURRENT RESULTS.
  ? 0


******************************************************
* ****************************************************** *
* *                                              * *
* *         CURRENT SECTION INFORMATION          * *
* *         ------- ------- -----------          * *
***                                              ***
        ROOF SLAB DEPTH= 170 MM
SLAB DEPTH AT FLOOR 3 = 170 MM
SLAB DEPTH AT FLOOR 2 = 170 MM
SLAB DEPTH AT FLOOR 1 = 170 MM

RIB DIMENSIONS
--- ----------

   ROOF LEVEL RIB : D= 350 MM   B= 250 MM
                 AST= 1.8248 %   ASC= 0 %
FLOOR 3 LEVEL RIB : D= 500 MM   B= 250 MM
                 AST= 2.02536 %   ASC= 0 %
FLOOR 2 LEVEL RIB : D= 500 MM   B= 250 MM
                 AST= 2.12334 %   ASC= 0 %
FLOOR 1 LEVEL RIB : D= 500 MM   B= 250 MM
                 AST= 2.16827 %   ASC= 0 %

EXTERNAL COLUMN DIMENSIONS
-------- ------ ----------

TOP STOREY COLUMNS : D= 200 MM   B= 300 MM
                 ASC= 1 %
   STOREY 3 COLUMNS : D= 200 MM   B= 300 MM
                 ASC= 1 %
   STOREY 2 COLUMNS : D= 250 MM   B= 300 MM
                 ASC= 1 %
   STOREY 1 COLUMNS : D= 250 MM   B= 300 MM
                 ASC= 2.5 %

INTERNAL COLUMN DIMENSIONS
-------- ------- ----------

TOP STOREY COLUMNS : D= 250 MM·  B= 300 MM
                 ASC= 1 %
   STOREY 3 COLUMNS : D= 250 MM   B= 300 MM
                 ASC= 2.25 %
   STOREY 2 COLUMNS : D= 450 MM   B= 300 MM
                 ASC= 1 %
   STOREY 1 COLUMNS : D= 450 MM   B= 300 MM
                 ASC= 3 %
***                                              ***
* *                                              * *
* ****************************************************** *
******************************************************


IF RESULTS ARE ACCEPTABLE TYPE 1 ELSE 0 ? 0
```

```
HOW MANY BEAMS ARE TO BE MODIFIED ? 4

SPECIFY FLO OR LEVEL,RIB DEPTH (MM) AND RIB WIDTH (MM)
  ? 4,325,250
  ? 3,450,250
  ? 2,450,250
  ? 1,450,250

HOW MANY EXTERNAL COLUMNS ARE TO BE MODIFIED ? 0

HOW MANY INTERNAL COLUMNS ARE TO BE MODIFIED ? 0
IF FRAME TO BE REANALYSED TYPE 1 ELSE 0 ? 1


FOR PROVISIONAL RESULTS OR FULL RESULTS TYPE 1 OR 0 ? 1


*****************************************************
* ************************************************* *
* *                                               * *
* *           PROVISIONAL  DESIGN  RESULTS         * *
***           ----------- ------ -------         ***

BEAM 4 ***CURRENT SECTION DIMENSIONS ACCEPTABLE***

BEAM 3 ***CURRENT SECTION DIMENSIONS ACCEPTABLE***

BEAM 2 ***CURRENT SECTION DIMENSIONS ACCEPTABLE***

BEAM 1 ***CURRENT SECTION DIMENSIONS ACCEPTABLE***



***                                               ***
* *                                               * *
* *                                               * *
* ************************************************* *
*****************************************************

EITHER - TYPE 1 - AND MODIFY SECTIONS
    OR - TYPE 0 - TO OBTAIN FULL PRINTOUT OF
                  CURRENT RESULTS.
  ? 0



*************************************************
* ******************************************** *
* *                                          * *
* *         CURRENT SECTION INFORMATION       * *
* *         ------- ------- -----------       * *
***                                          ***
        ROOF SLAB DEPTH= 170 MM
SLAB DEPTH AT FLOOR 3 = 170 MM
SLAB DEPTH AT FLOOR 2 = 170 MM
SLAB DEPTH AT FLOOR 1 = 170 MM

RIB DIMENSIONS
--- ----------

   ROOF LEVEL RIB : D= 325 MM   B= 250 MM
                    AST= 2.07393 %   ASC= 0 %
FLOOR 3 LEVEL RIB : D= 450 MM   B= 250 MM
                    AST= 2.41468 %   ASC= .139966 %
FLOOR 2 LEVEL RIB : D= 450 MM   B= 250 MM
                    AST= 2.49077 %   ASC= .231899 %
```

FLOOR 1 LEVEL RIB : D= 450 MM   B= 250 MM
                  AST= 2.52358 %   ASC= .27155 %

EXTERNAL COLUMN DIMENSIONS
-------- ------ ----------

TOP STOREY COLUMNS : D= 200 MM   B= 300 MM
                ASC= 1 %
   STOREY 3 COLUMNS : D= 200 MM   B= 300 MM
                ASC= 1 %
   STOREY 2 COLUMNS : D= 250 MM   B= 300 MM
                ASC= 1 %
   STOREY 1 COLUMNS : D= 250 MM   B= 300 MM
                ASC= 2.75 %

INTERNAL COLUMN DIMENSIONS
-------- ------ ----------

TOP STOREY COLUMNS : D= 250 MM   B= 300 MM
                ASC= 1 %
   STOREY 3 COLUMNS : D= 250 MM   B= 300 MM
                ASC= 2.25 %
   STOREY 2 COLUMNS : D= 450 MM   B= 300 MM
                ASC= 1 %
   STOREY 1 COLUMNS : D= 450 MM   B= 300 MM
                ASC= 3 %
***   ***
* *   * *
* ** *
****

IF RESULTS ARE ACCEPTABLE TYPE 1 ELSE 0 ? 1

RUNNING TIME: 19.5 SECS   I/O TIME :   60.8 SECS

Chapter 6

An Approach to the Design of Rigid Steel Frames

6.1 Introduction

The name of the program which is described in this chapter is SFD1. It was written in Algol, a more sophisticated programming language than BASIC, to carry out the elastic design of plane, rigid steel frames. Whilst the program itself is not presented here its structure is discussed in sufficient detail (when taken in conjunction with the information in Chapter 4) for similar programs to be written in whatever programming language the reader is proficient. SFD1 was written specifically to investigate the feasibility of an automatic design approach to the design of structural steelwork using standard sections. With this object in mind the scope of the program was limited to that of handling multi-storey, multi-bay frames comprising a regular pattern of rectangular cells.

The design method discussed here is not solely restricted to frames fabricated from structural steelwork. A similar approach could be adopted for structures fabricated from any material which may be assumed to behave in a linear elastic manner and which is available in a finite number of sections whose shape characteristics are known.

Whilst the manual design of rigid steel frames is essentially an iterative procedure, experienced designers will often approach their solutions by way of intuitive shortcuts. The precise nature of design experience is often difficult to analyse. It springs from an ability to form intuitive judgements concerning the probable effects of a number of variable parameters—a process which is often impossible to translate into a compact set of logical instructions. For this reason the automatic design based program which will be described is founded upon a formal iterative procedure, one which has been found to converge to an acceptable solution from even quite inept programmed assessments of starting values for the second moments of area.

6.2 The Design Method

The design method is illustrated by the simple flow diagram given in Fig. 6.1. With the exception of the input of data and the output of results the program

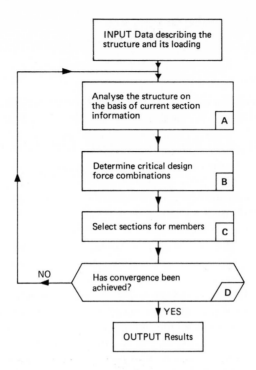

Figure 6.1 Flow diagram for program SFD1

consists of four separate and distinct stages linked together in a design loop. These stages are: the structural analysis; the search for critical force action combinations; the adoption of sections; and finally, a check to determine whether the criterion for design acceptance has been fulfilled. If the program is developed in this order, and each stage is proved to work before another one is added, then an apparently complex problem is greatly simplified. The first two stages, which are common to all frame designs, are discussed in Chapter 4. The last two stages are discussed in this chapter.

Before the design calculation may proceed, in addition to requiring data which specifies the frame geometry and its loading the program must also be furnished with assumed initial values for the relative second moments of area of the members. The influence of these assumed values on the course of a solution and its final result is discussed in Section 6.5.

When the design parameters have been input the computer calculation enters the loop $A \to B \to C \to D$ (see Fig. 6.1) in which it remains throughout an indeterminate number of iterations (each of which is based upon the better approximation suggested by the previous design iteration) until the process terminates when the criteria for convergence have been satisfied.

6.3 Programming for the Selection of Structural Sections
from a Standard List

6.3.1 Introduction

A facility offered by many computer systems is that of holding blocks of data in the computer filestore for subsequent retrieval and use at the time a relevant program is run. In a design context it is useful to store data of a permanent character (design tables, section properties, etc.) in this way. The data are then readily available for use, in whole or in part, with any pertinent program.

Ideally the section property lists which are thus held on file should be a complete record of published section information. For use with Program SFD1 the section property data are stored in ten one-dimensional arrays, each array holding a particular section property—the nominal depth, nominal breadth, weight per unit length, actual depth, actual breadth, web thickness, flange thickness, area, I_x and I_{yy} —for the whole range Universal Beam and Column sections.

Practical considerations, a desire to restrict section choice to a limited number of nominal sizes or the non-availability of some sections, usually dictate that fewer sections are made available for design purposes than are actually held permanently on file. In a similar way to that described in Section 2.2.4.2 for imposing a limited choice on reinforcement diameters, information concerning unwanted sections may be suppressed by using a 'preferred sections' array to transfer information from the filestore to the working file. Thus, if NPS is the number of preferred sections, then within the working file a section property array takes the form SECPROP (1 : NPS)—i.e the array will have NPS elements. In Program SFD1 the SECPROP( ) arrays hold section information in the order of increasing nominal depth, and within each grouping, increasing weight.

In published form Universal Beam and Column sections are classified in groups, the sections within a given group having as a common factor the same nominal overall dimensions. But whilst within each group the sections are tabulated in order of weight, it often happens that a group contains one or more sections of greater weight than those held in other groups having larger nominal dimensions. The classification is a logical one which presents few problems if the computer's choice of sections is restricted to those within a single group, or a number of groups whose weights do not overlap. If weight overlapping between groups is allowed then the procedure for finding the lightest section to satisfy programmed design criteria becomes more complex. Two ways of solving this problem are outlined below and discussed in greater detail in Sections 6.2.1.1 and 6.2.1.2.

One method, called Selection Procedure (A), works directly with the list of sections in their published order. Beginning with the first section in the list, each is considered in turn until one is found which satisfies all the programmed design criteria. A further search is then made through the list to ascertain whether an even lighter section is available. A disadvantage of the method is that the whole list must be examined to ensure that the lightest section is chosen. This time consuming activity may be speeded up by suppressing from the working section list those sections which, on the basis of experience, are clearly too small.

The second method, Selection Procedure (B), is one which works indirectly with the list of sections through the medium of a further array, Z( ) say, which is created to hold, in the order of *ascending weight*, the *location* of each section in the working list. By referring to the contents of Z( ) it is therefore possible to enter the section list at any level and to move up or down it in the knowledge that the direction of travel is indeed related to increasing or decreasing section weight.

In addition to published section information an array holding (of necessity, frequently updated) information concerning the cost per unit length of each section in the permanent list allows, in terms of material, a 'least cost' solution to be sought. For such solutions it would be arranged for array Z( ) to hold in order of *increasing cost* the location of each section in the working list.

Whichever section choice procedure is preferred two further facilities should be incorporated. One of these is concerned with translating a 'least weight' or a 'least cost' solution into a more practical arrangement of members by forcing preselected groups of members to take the properties of the largest section in the group (*cf*. Section 5.2.3.8). At the conclusion to each iteration the properties of these groups of members are the ones on which a further analysis is based. Thus to some extent the structural force actions become conditioned to assuming values which more readily encourage the use of 'same-section' groups.

The second facility is one which allows the formal section choice procedure to be bypassed so that check calculations may be performed upon designer-specified sections. When this facility is used the design procedure becomes a one-shot operation, the outcome of which is a list of 'Accept' or 'Fail' messages for each member in the structure. Incorporating this facility allows design checks to be made upon existing structures. It also allows designers of the 'pause and reflect' persuasion to use an otherwise automatic design based program in what is essentially a wholly decision design context.

6.3.1.1 Selection Procedure (A)

The flow diagram for this procedure is shown in Fig. 6.2. A structural member and its associated force actions are identified by assigning integer values to the variables I and J (see Section 4.2.1).

Section property arrays have the same length as the number of preferred sections (NPS). When seeking a least-weight solution the property array of immediate interest is the one which holds the weight per unit length of each section, called here WS( ). An integer variable called U defines the location of the current trial section in a section property array. Thus when U = N the section weight, the value of I_{xx} and the cross-sectional area of the Nth section

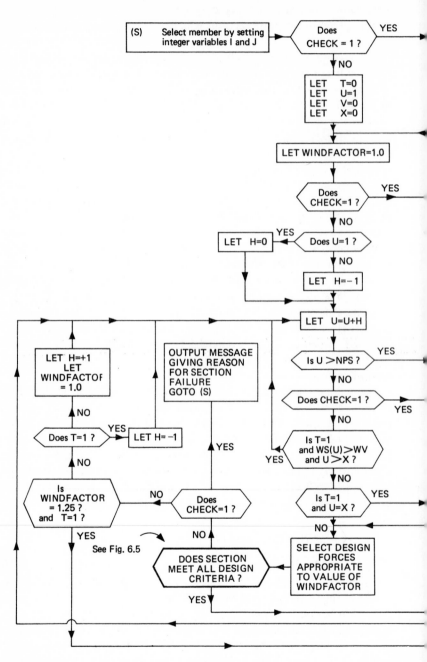

Figure 6.2 Flow diagram for Selection Procedure (A)

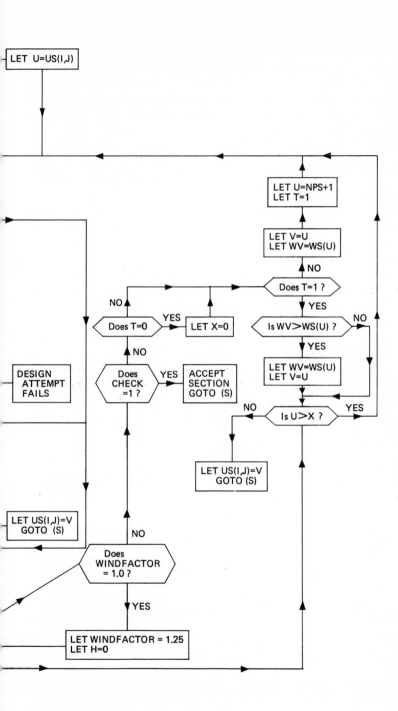

LET U=US(I,J)

LET U=NPS+1
LET T=1

LET V=U
LET WV=WS(U)

Does T=1 ?   NO

NO   Does T=0   YES   LET X=0   Is WV>WS(U) ?   NO

YES

NO   LET WV=WS(U)
LET V=U

DESIGN
ATTEMPT
FAILS

Does
CHECK
=1 ?   YES   ACCEPT
SECTION
GOTO (S)

NO   Is U>X ?   YES

LET US(I,J)=V
GOTO (S)

LET US(I,J)=V
GOTO (S)

NO   Does
WINDFACTOR
= 1.0 ?

YES

LET WINDFACTOR = 1.25
LET H=0

in the preferred sections list are defined by WS(N), IXX(N) and AS(N) respectively.

An integer variable called H1, which can take the values of $+1$, -1 or 0, determines whether the search proceeds up or down the list or is temporarily arrested so that further checks may be carried out on a section.

When the first section is found which meets all design criteria then the corresponding value of U is recorded in the variable called X. This value of U is also recorded in V and the weight of the section in WV. Whilst the values of V and WV may subsequently be altered if even lighter sections are found, the content of X remains the same throughout the procedure—its function being to define the lower boundary of the region in the section list still to be searched.

The dimensions of array US( )will be governed by the type of member which is to be designed. It will have the same dimensions as those set aside to store information concerning beam or column member properties. In this selection procedure the main purpose of US( ) is to record, at the end of each design iteration, current values of V for each member. In this way the member section properties required for the next analysis are readily identified. For example, if for a member which is defined by I and J the Nth section in the list is adopted, then: $V = N$, $US(I, J) = N$ and the second moment of area of the member is therefore $IS(I, J) = IXX(US(I, J))$.

If a simple check design calculation is being made, then the variable called CHECK is set equal to 1 and integers which represent the location in the section list of each assumed section are read into US( ) as data.

At appropriate stages in the selection procedure the allowable stresses are multiplied by a variable called WINDFACTOR which will take the value of 1.25 or 1.0 depending upon whether or not wind loading must be taken into account.

Two separate search operations are entailed in adopting a section. The value which is automatically assigned to the variable called T (either 0 or 1) defines each of these operations and dictates the route which will be taken through the flow diagram. For $T = 0$ the search begins with the first section in the list. With the WINDFACTOR set equal to 1.0, sections are considered in turn and rejected if they are unable to meet the 'no-wind' design criteria. On meeting a section which does satisfy these requirements the WINDFACTOR coefficient is set equal to 1.25; this affects both the value of the allowable stresses and the choice of design force actions. The section is now checked against the new requirements. If it is able to satisfy them then it is provisionally accepted by recording its value of U in the variables V and X and its unit weight in WV. Otherwise further sections are tested by advancing the value of U. If at any stage U takes a value greater than NPS then there is no suitable section for that member; the design calculation therefore ends and a diagnostic failure message is printed out.

If in the first search operation a section is provisionally chosen, then T becomes equal to 1 and a new search is directed from the top of the list downwards. In this case only sections having a unit weight less than that currently held in

WS( ) | 20 | 30 | 40 | 50 | 38 | 45 | 60 | 35 | 65 | 70 | NPS = 10
CHECK = 0

Fig. 6.3a

V	0	0	0	0	3	3	3	3	3	3	8	8	8	8	8	
T	0	0	0	0	0	1	1	1	1	1	1	1	1	1	1	
U	1	2	2	3	3	11	10	9	8	8	7	6	5	4	3	
X	0	0	0	0	3	3	3	3	3	3	3	3	3	3	3	
WIND FACTOR	1.0	1.0	1.25	1.0	1.25	1.0	1.0	1.0	1.0	1.25	1.0	1.0	1.0	1.0	1.0	
H	0	1	0	1	0	−1	−1	−1	−1	0	−1	−1	−1	−1	−1	
WV					40	40	40	40	40	40	35	35	35	35	35	
WS(U)	20	30	30	40	40	40	70	65	35	35	60	45	38	50	40	
US(I, J)					3	3	3	3	3	3	3	3	3	3	3	8

Fig. 6.3b

Figure 6.3 Array WS( ) and values taken by associated variables in Selection Procedure (A)

WV are tested. If and when one of these sections is found to be acceptable then the variables V and WV take the new values. This second search operation ends when $U = X$. The current value of V is then recorded in US(I, J).

A ten-element array WS( ) is shown in Fig. 6.3a. It is assumed to record the unit weights of sections belonging to three overlapping groups. If for a given member the second section in the list satisfies the 'no-wind' design criteria, and the third and eighth sections meet all criteria, then it is left to the reader to confirm that the variables used in the selection procedure take the values shown in Fig. 6.3b, and in particular that the eighth section in the list is the one which is finally recorded in US(I, J).

6.3.1.2 Selection Procedure (B)

Before entry into Selection Procedure (B) it is necessary to rearrange the list of structural steel section unit weights (which are held in array WS( )) so that they are in a more manageable form than offered by the published section list. This may be effected in the following steps:

1. Copy the contents of WS( ) into a similar array called SWS( ). Employ this to create an array called Z( ) which holds, in order of ascending weight, the location of each section in the preferred section list;

2. Replace the content of each element in array US( ), which initially holds a set of guessed sections as specified by their locations in the preferred sections list, with the locations of these sections in array Z( ).

By presenting the section unit weights to the computer in this way a more localised search through the section list may be made than was possible by Selection Procedure (A). The information needed for arrays Z( ) and US( ) could of course be obtained from a visual inspection of the list of section unit weights prior to it being presented to the computer as numerical data. But as a general rule, if the rearrangement of information can be reduced to a mechanical process, then it is better to relieve the program user of this task.

Since it is only necessary to execute these steps once in any program run, their natural position within the program is at a point immediately after the section information has been transferred from the filestore to the working file.

In BASIC language this section of program could take the form given below:

```
200 FOR P = 1 TO NPS
210 LET SWS (P) = WS(P)
220 NEXT P
230 FOR P = 1 TO NPS
240 LET E = 1000
250 FOR Y = 1 TO NPS
260 IF SWS(Y) = 0 THEN 320
270 IF E = SWS(Y) THEN 320
280 IF E − SWS(Y) < 0 THEN 320
290 LET E = SWS(Y)
300 LET Z(P) = Y
310 LET Q = Y
320 NEXT Y
330 LET SWS(Q) = 0
340 NEXT P
350 FOR I = 1 TO S
360 FOR J = 1 TO B + 1
370 FOR Y = 1 TO NPS
380 IF Z(Y) > < US(I, J) THEN 400
390 LET US(I, J) = Y
400 NEXT Y
410 NEXT J
420 NEXT I
```

The reader may confirm for himself that if NPS = 10 and the array WS( ) takes the following form:

| 20 | 30 | 40 | 50 | 38 | 45 | 60 | 35 | 65 | 70 |

then at lines 200 to 220 above, the array SWS( ) also becomes:

20	30	40	50	38	45	60	35	65	70

The array Z( ) is created at lines 230 to 340 and in this example it will take the form:

1	2	8	5	3	6	4	7	9	10

For a 3-bay, 4-storey frame (when relating to the columns) the original form of array US( ) might have been read into the computer as:

3	5	5	3
3	5	5	3
4	7	7	4
4	7	7	4

and at lines 350 to 420 the contents of this array would be transformed into:

5	4	4	5
5	4	4	5
7	8	8	7
7	8	8	7

The flow diagram which describes Selection Procedure (B) is shown in Fig. 6.4. The fact that in this procedure an entry may be made into the section list at any level means that array US( ) plays an active role from the beginning of the design calculation. The values of U which are read initially into US( ) (and subsequently transformed into their array Z( ) locations) may represent an informed guess concerning the sections of all members in the structure. This selection procedure makes it possible to check these sections immediately

208

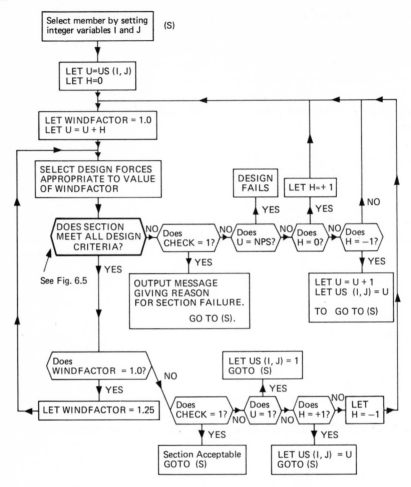

Figure 6.4 Flow diagram of Selection Procedure B

without the necessity of having to examine all the lighter sections in the list.

Assuming an entry point for U greater than unity then the section which is represented by that value of U is checked first. If it is found to satisfy all the design criteria, then H becomes equal to -1 and successively lighter sections are tested until one is met which fails to meet some requirement; in which case the next larger section than this one is accepted. If the first section which is tested fails to meet some requirement, then heavier sections are tried in succession (i.e. $H = +1$) until one of them is found to be acceptable.

In both of these situations the value of U which is recorded in US( ) at the end of a successful search represents a section whose position in the preferred section list is defined by the contents of Z(U). Therefore when a member is defined by integer variables I and J, $US(I, J) = U$ and the second moment of area of that member is $IS(I, J) = IXX(Z(U))$.

6.4 Member Design

6.4.1 Introduction

The primary role of an individual structural member (i.e. whether it is that of a 'beam' or a 'column') can often be recognized from a visual inspection of the loaded structure. The need to discriminate between member types arises mainly from the fact that 'beam' sections are usually restricted to those found in the Universal Beam list whilst 'column' sections might be chosen from either the Universal Column list or the Universal Beam list according to the relative importance of the axial load. It is for the designer to decide from which of the lists the column sections will be chosen. By setting a switch in the program to an appropriate value the preferred list of sections will be consulted. Otherwise there is little need for classification. Since most continuous frame members are subjected to a combination of bending moment and axial load, and within a programmed calculation it is difficult to assess the relative importance of each of these force actions, it is expedient to process both member types within the same formal design procedure.

When the role of members cannot easily be recognized then it is possible, at the expense of extra computing time, to program for this information to be output. The procedure in this case would be to locate, within each overall design iteration, both a Universal Beam section and a Universal Column section for each member in the structure. The properties of the lightest of these alternative sections would then be adopted to provide the basis for the next design iteration. The outcome of this preliminary investigation would be to identify the members (by virtue of the fact that a Universal Column section had been chosen) in which the axial force action had dominated section choice. The practicability of such a design—assessed by looking at the compatability of the section dimensions of members which frame into common joints, the number of different sections used in the structure, etc.—would then suggest the constraints necessary in a final design to produce an acceptable solution.

In structural steelwork the effective lengths of columns and compression flanges are matters for individual judgement rather than calculation. For this reason it is generally preferable to leave their assessment to the program user. Effective length coefficients should therefore be treated as necessary structural data which is to be read into the computer at the information input stage. But when a program is written to process a particular structural type which has familiar column end restraint and lateral beam restraint conditions, then it is reasonable to build conservative estimates of effective length into the program and to allow the designer the opportunity of overriding them should better estimates be available.

Problems connected with the choice of Steel Grade—whether more than one grade should be used, and if so, how they should be distributed throughout the structure—are the province of the designer. This information would also form part of the data input.

Setting aside possible modifications to the structural geometry, the many solutions implied by simultaneously varying the steel grade, discriminating between the two possible section types for columns and specifying 'same-section' groups of members makes it unreasonable to vary these parametrs automatically. So that even though a program may produce individual solutions on an automatic design basis, in a very real sense the route to an acceptable solution still rests with decisions made by the designer.

6.4.2 Programming for the Design of Members

By specifying the following additional data for each member in the structure their design may follow a common programmed procedure:

1. Coefficients which describe the effective column lengths of the member about both the $x-x$ and $y-y$ axes;
2. Coefficients which describe the compression flange length;
3. The chosen steel grade;
4. Whether the preferred section is to be taken from the Universal Beam or the Universal Column list;
5. Whether the maximum deflection within the span of the member is a critical consideration.

The blocks of data represented by items 1 to 5 above are stored in arrays which have the same dimensions as those of other member property arrays. And whilst this data will usually be assessed by the designer and read into the computer at the information input stage, standard structural cases may be catered for by automatically generating the data, in whole or in part, within the body of the program.

The member design procedure is a straightforward matter of programming a set of design requirements (in the case of Program SFD1 those of BS449: Part 2:1969). Two ways of selecting and adopting a section were discussed in Section 6.2. The flow diagrams describing these procedures (see Figs. 6.2 and 6.4) both include a decision box which contains the question, 'DOES THE SECTION MEET ALL DESIGN CRITERIA?' This box, together with the entry into it and the exits from it, is reproduced for reference in Fig. 6.5a. The design steps implied by the question which was posed have been expanded into the form of the detailed flow diagram shown in Fig. 6.5b. It has eight exits labelled (A); taken together they are equivalent to the single exit (A) in Fig. 6.5a. The exits labelled (B) in both figures are also equivalent.

With the exception of one check which compares the actual and allowable values of transverse deflection, the remaining design stages all consist of a comparison between an induced stress and an allowable stress. The values of induced stress are all obtained from calculations which are based upon assumed elastic behaviour and a knowledge of the section properties and maximum force actions. Each category of allowable stress is stored in its own array; a particular required value is identified either by some appropriate unique

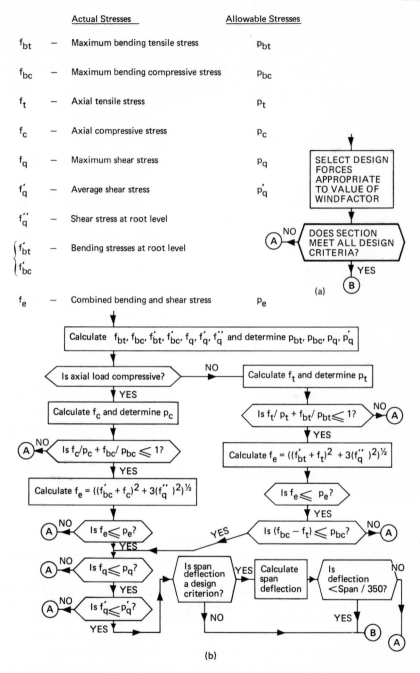

Actual Stresses | Allowable Stresses

f_{bt} — Maximum bending tensile stress p_{bt}

f_{bc} — Maximum bending compressive stress p_{bc}

f_t — Axial tensile stress p_t

f_c — Axial compressive stress p_c

f_q — Maximum shear stress p_q

f'_q — Average shear stress p'_q

f''_q — Shear stress at root level

$\begin{cases} f'_{bt} \\ f'_{bc} \end{cases}$ — Bending stresses at root level

f_e — Combined bending and shear stress p_e

SELECT DESIGN FORCES APPROPRIATE TO VALUE OF WINDFACTOR

DOES SECTION MEET ALL DESIGN CRITERIA?

(a)

Calculate f_{bt}, f_{bc}, f'_{bt}, f'_{bc}, f_q, f'_q, f''_q and determine p_{bt}, p_{bc}, p_q, p'_q

Is axial load compressive?

Calculate f_t and determine p_t

Calculate f_c and determine p_c

Is $f_t/p_t + f_{bt}/p_{bt} \leqslant 1$?

Is $f_c/p_c + f_{bc}/p_{bc} \leqslant 1$?

Calculate $f_e = ((f'_{bt} + f_t)^2 + 3(f''_q)^2)^{1/2}$

Calculate $f_e = ((f'_{bc} + f_c)^2 + 3(f''_q)^2)^{1/2}$

Is $f_e \leqslant p_e$?

Is $f_e \leqslant p_e$?

Is $(f_{bc} - f_t) \leqslant p_{bc}$?

Is $f_q \leqslant p_q$?

Is span deflection a design criterion?

Calculate span deflection

Is deflection < Span / 350?

Is $f'_q \leqslant p'_q$?

(b)

Figure 6.5 Flow diagram for the design of a structural steelwork member

quantity, or more generally by a value which is interpolated between the contents of neighbouring elements.

6.5 Initial Relative I-values: Their Effect on the Quality of an Automatic Solution and its Convergence

6.5.1 Introduction

In an indeterminate structure the magnitudes of the force actions due to continuity are a function of the relative values of the second moments of area of its members. The section properties, force actions and induced stresses are therefore interdependent. The primary function of structural design is to proportion the section of each member in such a way that the induced stresses and deformations are close to allowable limits without exceeding them; one way of achieving this state is to refine a guessed solution. It will be shown later that even though the rate of convergence is affected by the quality of the initially guessed solution, acceptable designs still emerge from quite inept assessments.

Consider, for example, the design of a symmetrical, rigid single bay frame. As a starting point for this design let us assume that the relative beam/column second moment of area ratio is $10^6 : 1$. That this is even a remotely possible solution is demonstrably untrue; the ratio of the I-values of the largest beam and the smallest column in the section list is only $570 : 1$ —and this is still a most unlikely combination of sections. Although account of continuity is taken, a first analysis of the frame gives force actions which are substantially those for a beam simply supported on axially loaded columns. The members designed to withstand these force actions have a beam/column I-value ratio of approximately $30 : 1$. This single design iteration has therefore taken a useful step towards an acceptable solution. A further analysis based upon these more realistic I-values reflects the fact that continuity indeed exists. The now substantial support moments effectively reduce the original beam design moment and this in turn leads to the adoption of a lighter beam section. In contrast with this the column sections must be increased to take account of the moment which now acts in conjunction with the axial load. The consequent reduction in the beam/column I-value ratio brings the solution even closer. In this example convergence would occur after six or seven iterations with a final I-value ratio of approximately $1.5 : 1$.

When designing to limited stress levels it would be ideal if each critical point in the structure could actually attain its allowable stress, but the fact that finite steps in strength occur between neighbouring sections in a list makes this event unlikely. This condition does, however, help to accelerate convergence. Because members are always marginally understressed (since we work to boundary values which must not be exceeded rather than to ones which must be attained) it is probable that at some stage in the design process two consecutive sets of (marginally different) sections will induce practically the same set of force actions. In this event further iterations would continue to reproduce the second of these sets of sections, thus proving that convergence has been achieved.

More complex frames respond to the same design approach. Early design iterations affect the whole of the structure and during this period of flux there is a general tendency for members to assume their correct order of relative I-value. Following this settling down period the structure begins to respond to the design process as an assemblage of simpler frames. Even though a modification to the section size of a single member affects the values of the force actions throughout the whole structure the effect is often of only local significance. Because of this the sections of members remote from the source of such a disturbance will usually have a sufficient reserve of strength to preclude the need, at least for this reason, for change. As the course of the design progresses some regions in the structure will converge more rapidly to their final form than others. But throughout this transformation from an approximate to an accurate solution one or more of the member sections will continue to change until a stage is reached where all the sections are reproduced in two consecutive designs.

Whilst convergence may usually be defined in this way, at times the solution will oscillate and formal convergence is never quite attained. The climate necessary for this state occurs when the section of one of the members meeting at a joint satisfies current design requirements within close limits and simultaneously the other members to which it is connected are relatively understressed. If this happens at a time when the general tendency has been for the stiffness of the highly stressed member to decrease relative to the other members with which it frames, then reanalysis at this stage may show a sufficient increase in the end-moment of that member to necessitate the choice of the next larger section—and this without a corresponding change in the sections of the other members at the joint. In this situation a further analysis may then show that the stiffness of what was originally a highly stressed member has increased sufficiently to reduce the offending force action to a level which now merits a reduction in the size of the section. This alternating increase and decrease in a section size is a vicious circle from which there is no exit. But once the condition is recognized as a possibility it then becomes a simple matter to check upon alternate as well as consecutive designs to determine whether convergence has indeed been achieved.

The assumptions on which a designer/programmer makes his initial assessments of relative I-values follow naturally from considerations of structural behaviour. Not only does the change in the size of a section have little effect on the magnitude of the force actions in remote regions of the structure, but in this respect neither does the absolute size of the section itself have much significance. Relative I-values may therefore be assessed on a local basis without considering their relationship to other parts of the structure. Regardless of their location in the structure a beam and the columns into which it frames will usually have a relative I-value ratio in the range $1:1$ to $3:1$. The fact that similar member types drawn from two remote regions may themselves have relative I-values of 10:1 say, is irrelevant. Even though precise values cannot be ascribed without knowing the final solution, if realistic local relative values are initially assumed then convergence is rapid.

6.5.2 Examples of Computer Aided Design in Structural Steelwork

Details of the geometry and loading of two types of frame which have been designed by Program SFD1 are given in Fig. 6.6; some information concerning the outcome of these computer aided designs is given in Tables 6.1a and b. Apart from the fact that these structures are of general interest, they each pose a problem which it is essential that an automatic structural steelwork design program should be able to handle. The 8-storey frame is an example in which the wind load forces dominate the choice of the lower storey sections; the 3-bay frame is typical of structures having working areas serviced by relatively narrow corridors. In the latter case, because of the 'long–short–long' arrangement of bays the sections which are adopted for the corridor beams have a critical effect upon the stiffness of the internal joints, which affects in turn the sizes of the remaining members framing into those joints. In this situation it is possible that difficulty might be experienced in attaining convergence.

Each frame type was designed on the basis of six different sets of initial beam/ column I-value ratios, viz. $10^{-6}:1$, $1:1$, $2:1$, $3:1$, $10:1$ and $10^6:1$. The effect of a further set of conditions was examined in the case of the 8-storey frame (in Table 6.1a this is referred to as the 'unsymmetrical' case). There it was assumed that for relative beam I-values of 1 the left hand and right hand columns would have initial values of 10^6 and 10^{-6} respectively. This is equivalent to assuming that the structure will behave like a stack of eight propped cantilevers.

The section list held 45 Universal Beam sections and 27 Universal Column sections; in each frame the vertical members were assumed to behave primarily as axial-load carrying members and as such their choice of section was restricted to those in the Universal Column list. For each set of initial assumptions two designs were carried out; one on the basis of a free choice of sections from within the appropriate list and the other was constrained to give 'same-section' groups of members.

The most noteworthy design is the one which is based upon assumed unsymmetrical behaviour and a free choice of sections (see Table 6.1a). Structural symmetry is never forced upon this solution but it evolves naturally nevertheless. Within seven iterations the design converges to precisely the same solution as that given by four others, each based upon widely differing initial assumptions.

To summarize the course of this solution, at significant stages in the design average beam/column I-values have been taken across the whole frame. The first design iteration converts starting values of

$$10^6 \longmapsto\!\!\!\!\!\!\!\!\!\overset{1}{}\!\!\!\!\!\!\!\!\!\longmapsto 10^{-6} \qquad \text{to} \qquad 1.0 \longmapsto\!\!\!\!\!\!\!\!\!\overset{1.0}{}\!\!\!\!\!\!\!\!\!\longmapsto 0.1$$

At the end of the fourth iteration symmetry emerges with values of

$$0.58 \vdash\!\!\!\!\!\overset{\textstyle 1.0}{\rule{3cm}{0.4pt}}\!\!\!\!\!\dashv 0.58$$

Three further iterations produce convergence with final values of

$$0.61 \vdash\!\!\!\!\!\overset{\textstyle 1.0}{\rule{3cm}{0.4pt}}\!\!\!\!\!\dashv 0.61$$

This is equivalent to a beam/column I-value ratio of 1.64:1.

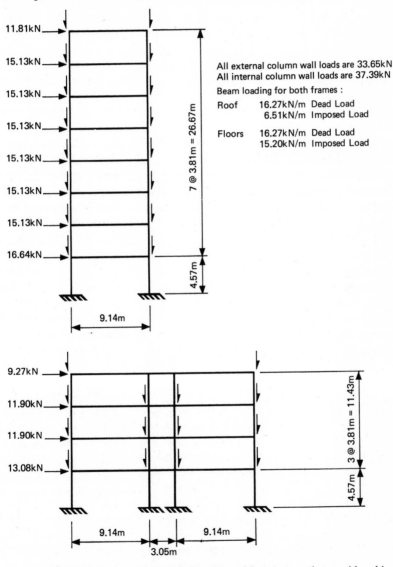

All external column wall loads are 33.65kN
All internal column wall loads are 37.39kN

Beam loading for both frames :

Roof   16.27kN/m Dead Load
              6.51kN/m Imposed Load

Floors   16.27kN/m Dead Load
           15.20kN/m Imposed Load

Figure 6.6 Structural geometry and loading of frame examples considered in Section 6.4.2

Table 6.1a 8-Storey, 1-Bay Frame

FREE CHOICE OF SECTIONS			'SAME-SECTION' GROUPS		
Assumed Relative I_B/I_C	Required Number of Iterations	Weight of Frame (Kgs)	Assumed Relative I_B/I_C	Required Number of Iterations	Weight of Frame (kgs)
10^{-6}:1	5	13177	10^{-6}:1	4	15357
1:1	3	13170	1:1	3	15357
2:1	3	13052	2:1	3	15357
3:1	4	13052	3:1	4	15357
10:1	6	13052	10:1	5	15243
10^6:1	8	13052	10^6:1	7	15243
Unsymmetrical Case	7	13052	Unsymmetrical Case	4	15357

Table 6.1b 4-Storey, 3-Bay Frame

FREE CHOICE OF SECTIONS			'SAME-SECTION' GROUPS		
Assumed Relative I_B/I_C	Required Number of Iterations	Weight of Frame (kgs)	Assumed Relative I_B/I_C	Required Number of Iterations	Weight of Frame (kgs)
10^{-6}:1	4	10605	10^{-6}:1	4	10914
1:1	5	10745	1:1	3	10893
2:1	5	10965	2:1	4	10766
3:1	5	10689	3:1	4	10766
10:1	7	10609	10:1	5	10891
10^6:1	8	10677	10^6:1	6	10766

The first iteration produced a frame weighing only 5% more, and the fifth iteration a frame weighing 1.5% less, than the final structural weight; intermediate designs give weights within these limits. From the outset, therefore, the total structural weight is sensibly constant and the main consequence of the design process is to redistribute the material more effectively. Table 6.2 records the unit weight of each member throughout the course of the solution; the chained lines in that table focus attention on the stage at which individual members attain their final section value. It is evident from this that the two top storeys settle down less rapidly than do the remainder of the structure.

By comparing the results quoted in Table 6.1a and b it can be seen that for both frame types initial beam/column I-value assessments in the range 1:1 to 3:1 lead to the most rapid solutions. Otherwise, if unreal assessments are to be made then it is better to err towards assuming fully fixed beams (i.e. a 10^{-6}:1 beam/column I-value ratio) than it is to assume the simply supported beam condition (i.e. a 10^6:1 beam/column I-value ratio). This follows from the fact

Table 6.2

Floor Level	Beams—Mass/metre (kgs)						
8	82	67	60	54	51	54	54
7	92	82	67	67	74	74	74
6	92	74	74	74	74	74	74
5	92	74	74	74	74	74	74
4	92	103	82	82	82	82	82
3	92	92	89	89	89	89	89
2	92	101	92	92	92	92	92
1	92	113	113	113	113	113	113

Storey Level	Left Hand Columns—Mass/metre (kgs)						
8	89	52	71	73	73	89	89
7	129	73	73	73	73	73	73
6	153	97	89	89	89	89	89
5	177	118	97	97	97	97	97
4	177	118	97	118	118	118	118
3	202	129	118	118	118	118	118
2	235	153	129	129	129	129	129
1	283	283	243	213	213	213	213

Storey Level	Right Hand Columns—Mass/metre (kgs)						
8	23	46	71	73	73	89	89
7	23	52	73	73	73	73	73
6	37	71	89	89	89	89	89
5	46	89	97	97	97	97	97
4	46	97	118	118	118	118	118
3	60	118	118	118	118	118	118
2	71	129	129	129	129	129	129
1	88	185	213	213 ·	213	213	213
Iteration Number	1	2	3	4	5	6	7

that the former condition is a better approximation to actual continuous frame behaviour than is the latter. Since it is unlikely that precise initial values can ever be assigned (and from a practical viewpoint it is relatively unimportant that they should need to be) a reasonable compromise is to build a blanket ratio of 2:1 into the program with the option of overriding this should a better assessment be available.

As might be expected, 'same-section' member group solutions tend to converge more rapidly than those resulting from a free choice of sections. This behaviour is particularly evident in the 'unsymmetrical' starting case (see Table 6.1a) where, after the first analysis, groups of four columns are coerced into taking symmetrical values.

Within each of three of the four groups of designs a significant number of the solutions converge to precisely the same solution in spite of widely divergent initial assessments; and in neither group does a total frame weight vary by more than 1.4% of the 'stable' solution in that group.

Computing time is a function of the efficiency of the computer used, the form of the program, the complexity of the frame and the number of iterations. Running Program SFD1 on an ICL 1904S computer one design iteration for the 8-storey, single bay frames (16 joints, 24 members, 9 analyses per iteration) took 29 seconds; the 4-storey, 3-bay frames (16 joints, 28 members, 13 analyses per iteration) took 40 seconds per iteration.

Index

220

222